Aging and Ethics

Contemporary Issues in Biomedicine, Ethics, and Society

Aging and Ethics, edited by *Nancy S. Jecker, 1991*

New Harvest: *Transplanting Body Parts and Reaping the Benefits,* edited by *C. Don Keyes,* in collaboration with coeditor *Walter E. Wiest, 1991*

Beyond Baby M, edited by *Dianne M. Bartels, Reinhard Priester, Dorothy E. Vawter,* and *Arthur L. Caplan, 1990*

Reproductive Laws for the 1990s, edited by *Sherrill Cohen* and *Nadine Taub, 1989*

The Nature of Clinical Ethics, edited by *Barry Hoffmaster, Benjamin Freedman,* and *Gwen Fraser, 1988*

What Is a Person?, edited by *Michael F. Goodman, 1988*

Advocacy in Health Care, edited by *Joan H. Marks, 1986*

Which Babies Shall Live?, edited by *Thomas H. Murray* and *Arthur L. Caplan, 1985*

Feeling Good and Doing Better, edited by *Thomas H. Murray, Willard Gaylin,* and *Ruth Macklin, 1984*

Ethics and Animals, edited by *Harlan B. Miller* and *William H. Williams, 1983*

Profits and Professions, edited by *Wade L. Robison, Michael S. Pritchard,* and *Joseph Ellin, 1983*

Visions of Women, edited by *Linda A. Bell, 1983*

Medical Genetics Casebook, by *Colleen Clements, 1982*

Who Decides?, edited by *Nora K. Bell, 1982*

The Custom-Made Child?, edited by *Helen B. Holmes, Betty B. Hoskins,* and *Michael Gross, 1980*

Medical Responsibility, edited by *Wade L. Robison* and *Michael S. Pritchard, 1979*

Contemporary Issues in Biomedical Ethics, edited by *John W. Davis, Barry Hoffmaster,* and *Sarah Shorten, 1979*

Aging and Ethics

Philosophical Problems in Gerontology

Edited by

Nancy S. Jecker

University of Washington School of Medicine, Seattle, Washington

Humana Press • Totowa, New Jersey

DEDICATION

For my mother,
Mae Anna Driscoll Silbergeld
November 23, 1923–November 22, 1990

Library of Congress Cataloging in Publication Data

Main entry under title:
Aging and Ethics: philosophical problems in gerontology / edited by Nancy S. Jecker.
p. cm. — (Contemporary issues in biomedicine, ethics, and society)
Includes index.
ISBN 0-89603-201-9, (hardcover)
ISBN 0-89603-255-8, (paperback)
1. Gerontology—Moral and ethical aspects. 2. Aging—Moral and ethical aspects. 3. Death—Moral and ethical aspects. I. Jecker, Nancy Ann Silbergeld. II. Series.
HQ1061.A4555 1991
305.26—dc20 91-10792
CIP

999 Riverview Dr., Suite 208
Totowa, NJ 07512

Printed in the United States of America. 10 9 8 7 6 5 4 3 2

Preface

The Aging Self and the Aging Society

Ethical issues involving the elderly have recently come to the fore. This should come as no surprise: Since the turn of the century, there has been an eightfold increase in the number of Americans over the age of sixty-five, and almost a tripling of their proportion to the general population. Those over the age of eighty-five—the fastest growing group in the country—are twenty-one more times as numerous as in 1900. Demographers expect this trend to accelerate into the twenty-first century.

The aging of society casts into vivid relief a number of deep and troubling questions. On the one hand, as individuals, we grapple with the immediate experience of aging and mortality and seek to find in it philosophical or ethical significance. We also wonder what responsibilities we bear toward aging family members and what expectations of others our plans for old age can reasonably include. On the other hand, as a community, we must decide: What special role, if any, do older persons occupy in our society? What constitutes a just distribution of medical resources between generations? And, How can institutions that serve the old foster imperiled values, such as autonomy, self-respect, and dignity?

Only recently have we begun to explore these themes, yet already a rich and fruitful literature has grown up around them. The essays in this volume address a wide range of concerns, incorporate a multiplicity of disciplines, and afford a broad perspective for viewing the many timely and important issues members of an aging society face. Somewhat procrusteanly, the issues dealt with here can be viewed at four levels. The first level is that of the individual experience of personal aging, anticipation of death, and understanding of finitude. Sally Gadow depicts these aspects vividly, describing the experience of living in an aging body and the reaction an aging body prompts from others. The essays by Baier and Moody explore the meaning of aging by calling attention to the whole-life perspective later years afford and the opportunity for self-knowledge and discovery this perspective creates. Thomas Cole's paper also treats the individual experience of aging, by elucidating its historical expression in the metaphor of life as a journey. The incompleteness of this motif for raising larger questions about the social significance of aging suggests the need to transcend personal aging and consider aging in a social context.

Moving then from a phenomenological and personal level to an interpersonal and public second level, Hagestad describes the change in family structure an aging society implies. English and Callahan attempt to say what exactly grown children owe their elderly parents, as distinguished from what it would be kind or salutary for them to provide. Fry considers concrete problems of medical decision making with elderly patients. She illustrates the significance of the elderly patient–adult offspring and elderly patient–nurse relationships for safeguarding autonomous choices. Finally, Barker and Jecker touch on the ways that patient–family relationships can enable (or disable) responsible medical decision making.

The third level is oriented less toward the intimate and interpersonal aspects of aging and more toward social consciousness of and collective responsibility for it. One issue that arises at this level is whether it is ever ethically appropriate to disenfranchise the elderly from publicly financed life-extending medical care. Callahan and Daniels suggest that it is, and delineate the conditions where limiting health care for older persons is consistent with standards of justice. Waymack rejects their arguments and attempts to show that using age as a criterion of exclusion from medical care is not ultimately defensible. Jecker likewise casts doubt on the proposal to exclude the elderly from life-extending medical care. She challenges the concept of a "natural life span" upon which this proposal rests. A related issue is how to define general health care goals for older Americans. Menzel shows that figuring the costs of meeting health care goals is enormously complex. Boyajian points out that a large gap exists between initially stated goals and their achievement in political and administrative processes.

A fourth and final level brings us full circle to the general philosophical level with which our inquiry began. Jonsen considers the salience of age and argues that the elderly occupy a privileged position by virtue of their collective past contributions. Schneiderman and Jecker ask, What is a fitting stance toward human finitude and aging? A final essay, by Nagel, suggests that coming to terms with our "cosmic unimportance" requires learning to feel at home with certain ironies in human existence.

Nancy S. Jecker

Acknowledgments

The Role of Intimate Others in Medical Decision Making, by Nancy S. Jecker originally appeared in *The Gerontologist,* vol 30, 1990, ©Gerontological Society of America

Limiting Health Care for the Old, by Daniel Callahan originally appeared in *Setting Limits,* Simon & Schuster, Inc., ©1987 by Daniel Callahan

Families as Caregivers: The Limits of Morality, by Daniel Callahan originally appeared in *The Archives of Physical Medicine and Rehabilitation,* vol 69, 1988

The Meaning of Life and the Meaning of Old Age, by Harry R. Moody originally appeared in *What Does It Mean to Grow Old?,* Durham, NC, Duke University Press 1986, Thomas R. Cole and Sally Gadow, eds

The Absurd, by Thomas Nagel originally appeared in the *Journal of Philosophy,* LXIII:20 (1971)

The Aging Society as a Context for Family Life, by Gunhild O. Hagestad was reprinted by permission of *Daedalus,* Journal of the American Academy of Arts and Sciences, "The Aging Society," Winter 1986, Vol. 115, No. 1, Cambridge, MA

What Do Grown Children Owe Their Parents?, by Jane English originally appeared in *Philosophical and Legal Reflections on Parenthood,* New York: Oxford University Press, 1979, Onora O'Neill and William Ruddick, eds.

Contents

v Preface
xi Contributors

The Aging Individual

3 The Meaning of Life
Kurt Baier
51 The Meaning of Life in Old Age
Harry R. Moody
93 Oedipus and the Meaning of Aging: *Personal Reflections and Historical Perspectives*
Thomas R. Cole
113 Recovering the Body in Aging
Sally Gadow

Aging and Filial Responsibility

123 The Aging Society as a Context for Family Life
Gunhild O. Hagestad
147 What Do Grown Children Owe Their Parents?
Jane English
155 Families as Caregivers: *The Limits of Morality*
Daniel Callahan
171 Health Care and Decision Making
Sara T. Fry
187 Rethinking Family Loyalties
Evelyn M. Barker
199 The Role of Intimate Others in Medical Decision Making
Nancy S. Jecker

Distributive Justice in an Aging Society

219 Limiting Health Care for the Old
Daniel Callahan
227 A Lifespan Approach to Health Care
Norman Daniels
247 Old Age and the Rationing of Scarce Health Care Resources
Mark H. Waymack
269 Appeals to Nature in Theories of Age-Group Justice
Nancy S. Jecker
285 Paying the Real Costs of Lifesaving
Paul T. Menzel
307 Intent and Actuality: *Sacrificing the Old and Other Health Care Goals*
Jane A. Boyajian

Philosophical Reflections on Aging and Death

341 Resentment and the Rights of the Elderly
Albert R. Jonsen
353 Ancient Myth and Modern Medicine: *Lessons from Baucis and Philemon*
Lawrence J. Schneiderman
367 The Meaning of Temporality in Old Age
Nancy S. Jecker
375 The Absurd
Thomas Nagel

389 Index

Contributors

KURT BAIER • *Department of Philosophy, University of Pittsburgh, Pittsburgh, PA*

EVELYN BARKER • *Department of Philosophy, University of Maryland, Catonsville, MD*

JANE BOYAJIAN • *Washington State Department of Social and Health Services, Olympia, WA*

DANIEL CALLAHAN • *Hastings Center, Briarcliff Manor, NY*

THOMAS COLE • *University of Texas Medical Branch, Institute for the Medical Humanities, Galveston, TX*

NORMAN DANIELS • *Department of Philosophy, Tufts University, Medford, MA*

JANE ENGLISH • *formerly, Department of Philosophy, University of North Carolina at Chapel Hill, Chapel Hill, NC*

SARAH FRY • *University of Maryland, School of Nursing, Baltimore, MD*

SALLY GADOW • *University of Colorado, School of Nursing, Denver, CO*

GUNHILD O. HAGESTAD • *Northwestern University, School of Education, Chicago, Ill*

NANCY S. JECKER • *Department of Philosophy, University of Washington, and University of Washington, School of Medicine, Seattle, WA*

ALBERT JONSEN • *University of Washington, School of Medicine, Seattle, WA*

PAUL MENZEL • *Department of Philosophy, Pacific Lutheran University, Tacoma, WA*

HARRY R. MOODY • *Brookdale Center on Aging, Hunter College, New York, NY*

THOMAS NAGEL • *Department of Philosophy, New York University, New York, NY*

LAWRENCE SCHNEIDERMAN • *University of California, School of Medicine, La Jolla, CA*

MARK WAYMACK • *Department of Philosophy, Loyola University, Chicago, IL*

The Aging Individual

The Meaning of Life

Kurt Baier

Tolstoy, in his autobiographical work, "A Confession," reports how, when he was fifty and at the height of his literary success, he came to be obsessed by the fear that life was meaningless.

> At first I experienced moments of perplexity and arrest of life, as though I did not know what to do or how to live; and I felt lost and became dejected. But this passed, and I went on living as before. Then these moments of perplexity began to recur oftener and oftener, and always in the same form. They were always expressed by the questions: What is it for? What does it lead to? At first it seemed to me that these were aimless and irrelevant questions. I thought that it was all well known, and that if I should ever wish to deal with the solution it would not cost me much effort; just at present I had no time for it, but when I wanted to, I should be able to find the answer. The questions however began to repeat themselves frequently, and to demand replies more and more insistently; and like drops of ink always falling on one place they ran together into one black blot.[1]

A Christian living in the Middle Ages would not have felt any serious doubts about Tolstoy's questions. To him it would have seemed quite certain that life had a meaning and quite clear what it was. The medieval Christian world picture assigned to humans a highly significant, indeed the central part in the grand scheme of things. The universe was made for the express purpose of providing a stage on which to enact a drama starring Man in the title role.

To be exact, the world was created by God in the year 4004 BC. Man was the last and the crown of this creation, made in the likeness of God, placed in the Garden of Eden on earth, the fixed center of the universe, round which revolved the nine heavens of the sun, the moon, the planets, and the fixed stars, producing as they revolved in their orbits the heavenly harmony of the spheres, and this gigantic universe was created for the enjoyment of humans, who were originally put in control of it. Pain and death were unknown in paradise, but this state of bliss was not to last. Adam and Eve ate of the forbidden tree of knowledge, and life on this earth turned into a death march through a vale of tears. Then, with the birth of Jesus, new hope came into the world. After He had died on the cross, it became at least possible to wash away with the purifying water of baptism some of the effects of Original Sin and to achieve salvation. That is to say, on condition of obedience to the law of God, humans could now enter heaven and regain the state of everlasting, deathless bliss, from which they had been excluded because of the sin of Adam and Eve.

To the medieval Christian, the meaning of human life was, therefore, perfectly clear. The stretch on earth is only a short interlude, a temporary incarceration of the soul in the prison of the body, a brief trial and test, fated to end in death, the release from pain and suffering. What really matters is the life after the death of the body. One's existence acquires meaning not by gaining what this life can offer, but by saving one's immortal soul from death and eternal torture, by gaining eternal life and everlasting bliss.

The scientific world picture that has found ever more general acceptance from the beginning of the modern era onwards is in profound conflict with all this. At first, the Christian conception of the world was discovered to be erroneous in various important details. The Copernican theory showed the earth as merely one of several planets revolving around the sun, and the sun itself was later seen to be merely one of many fixed stars, each of which was itself the nucleus of a solar system similar to our own. Humans, instead of occupying the center of creation, proved to be merely the inhabitants of a celestial body no different from millions of others. Furthermore, geological investigations revealed that the universe was not created a few thousand years ago, but was probably millions of years old.

Disagreements over details of the world picture, however, are only superficial aspects of a much deeper conflict. The appropriateness of the whole Christian outlook is at issue. For Christianity, the world must be regarded as the "creation" of a kind of Superman, a person possessing all the human excellences to an infinite degree and none of the human weaknesses, who has made humans in His image, feeble, mortal, foolish copies of Himself. In creating the universe, God acts as a sort of playwright-cum-legislator-cum-judge-cum-executioner. In the capacity of playwright, He creates the historical world process, including humans. He erects the stage and writes, in outline, the plot. He creates the *dramatis personae* and watches over them with the eye partly of a father, partly of the law. While on stage, the actors are free to extemporize, but if they infringe on the divine commandments, they are later dealt with by their creator in His capacity of judge and executioner.

Within such a framework, the Christian attitudes towards the world are natural and sound: it is natural and sound to think that all is arranged for the best even if appearances belie it; to resign oneself cheerfully to one's lot; to be filled with awe and veneration in regard to anything and everything that happens; to want to fall on one's knees and worship and praise the Lord. These are wholly fitting attitudes within the framework of the world view just outlined. This world view must have seemed wholly sound and acceptable because it offered the best explanation then available of all the observed phenomena of nature.

As the natural sciences developed, however, more and more things in the universe came to be explained without the assumption of a supernatural creator. Science, moreover, could explain them better, that is, more accurately and more reliably. The Christian hypothesis of a supernatural maker, whatever other needs it was capable of satisfying, was, at any rate, no longer indispensable for the purpose of explaining the existence or occurrence of anything. In fact, scientific explanations do not seem to leave any room for this hypothesis. The scientific approach demands that we look for a natural explanation of anything and everything. The scientific way of looking at and explaining things has yielded an immensely greater measure of understanding of, and control over, the universe than any other way. Also, when one looks at the world in this scientific way, there seems to be no room for a personal relationship between human beings and a supernatural perfect being ruling and guiding people. Hence, many scientists and educated men and women have come to feel that the Christian attitudes towards the world and human existence are inappropriate. They have become convinced that the universe and human existence in it are without a purpose and, therefore, devoid of meaning.[2]

The Explanation of the Universe

Such beliefs are disheartening and unplausible. It is natural to keep looking for the error that must have crept into our arguments, and if an error has crept in, then it is most likely to have crept in with science. For before the rise of science, people did not entertain such melancholy beliefs, whereas the scientific world picture seems literally to force them on us.

There is one argument that seems to offer the desired way out. It runs somewhat as follows. Science and religion are not really in conflict. They are, on the contrary, mutually complementary, each doing an entirely different job. Science gives provisional, if precise, explanations of small parts of the universe; religion gives final and overall, if comparatively vague, explanations of the universe as a whole. The objectionable conclusion, that human existence is devoid of meaning, follows only if we use scientific explanations where they do not apply, namely, where total explanations of the whole universe are concerned.[3]

After all, the argument continues, the scientific world picture is the inevitable outcome of rigid adherence to scientific method and explanation, but scientific, that is, causal, explanations from their very nature are incapable of producing real illumination. They can at best tell us *how* things are or have come about, but never *why*. They are incapable of making the universe intelligible, comprehensible, meaningful to us. They represent the universe as meaningless, not because it is meaningless, but because scientific explanations are not designed to yield answers to investigations into the why and wherefore, into the meaning, purpose, or point of things. Scientific explanations (this argument continues) began, harmlessly enough, as partial and provisional explanations of the movement of material bodies, in particular the planets, within the general framework of the medieval world picture. Newton thought of the universe as a clock made, originally wound up, and occasionally set right by God. His laws of motion only revealed the ways in which the heavenly machinery worked. Explaining the movement of the planets by these laws was analogous to explaining the machinery of a watch. Such explanations showed *how* the thing worked, but not *what it was for* or *why* it existed. Just as the explanation of how a watch works can help our understanding of the watch only if, in addition, we assume that there is a watchmaker who has designed it for a purpose, made it, and wound it up, so the Newtonian explanation of the solar system helps our understanding of it only on the similar assumption that there is some divine artificer who has designed and made this heavenly clockwork for some purpose, has wound it up, and perhaps even occasionally sets it right, when it is out of order.

Socrates, in the Phaedo, complained that only explanations of a thing showing the good or purpose for which it existed could offer a *real* explanation of it. He rejected the kind of explanation we now call "causal" as no more than mentioning "that without which a cause could not be a cause," that is, as merely a necessary condition, but not the *real* cause, the real explanation.[4] In other words, Socrates held that *all* things can be explained in two different ways: either by mentioning merely a necessary condition, or by giving the real cause. The former is not an elucidation of the *explicandum*, not really a help in understanding it, in grasping its "why" and "wherefore."

This Socratic view, however, is wrong. It is not the case that there are two kinds of explanations for everything, one partial, preliminary, and not really clarifying, the other full, final, and illuminating. The truth is that these two kinds of explanations are equally explanatory, equally illuminating, and equally full and final, but that they are appropriate for different kinds of *explicanda*.

When in an uninhabited forest we find what looks like houses, paved streets, temples, cooking utensils, and the like, it is no great risk to say that these things are the ruins of a deserted city, that is to say, of something manufactured. In such a case, the appropriate explanation is teleological, that is, in terms of the purposes of the builders of that city. On the other hand, when a comet approaches the earth, it is similarly a safe bet that, unlike the city in the forest, it was not manufactured by intelligent creatures and that, therefore, a teleological explanation would be out of place, whereas a causal one is suitable.

It is easy to see that in some cases causal, and in others teleological, explanations are appropriate. A small satellite circling the earth may or may not have been made by humans. We may never know which is the true explanation, but either hypothesis is equally explanatory. It would be wrong to say that only a teleological explanation can *really* explain it. Either explanation would yield complete clarity although, of course, only one can be true. Teleological explanation is only one of several that are possible.

It may indeed be strictly correct to say that the question *"Why* is there a satellite circling the earth?" can only be answered by a teleological explanation. It may be true that "Why?"—questions can really be used properly only in order to elicit *someone's reasons* for doing something. If this is so, it would explain our dissatisfaction with causal answers to 'Why?"—questions. However, even if it is so, it does not show that "Why is the satellite there?" *must be answered by a teleological explanation.* It shows only that either it must be so answered or it must not be asked. The question "Why have you stopped beating your wife?" can be answered only by a teleological explanation, but if you have never beaten her, it is an improper question. Similarly, if the satellite is not manufactured, "Why is there a satellite?" is improper, since it implies an origin it did not have. Natural science can indeed only tell us *how* things in nature have come about and not *why,* but this is so not because something else can tell us the *why* and *wherefore,* but because there is none.

There is, however, another point that has not yet been answered. The objection just stated was that causal explanations did not even set out to answer the crucial question. We ask the question "Why?" but science returns an answer to the question "How?" It might now be conceded that this is no ground for a complaint, but perhaps it will instead be said that causal explanations do not give complete or full answers even to that latter question. In causal explanations, it will be objected, the existence of one thing is explained by reference to its cause, but this involves asking for the cause of that cause, and so on, *ad infinitum.* There is no resting place that is not as much in need of explanation as what has already been explained. Nothing at all is ever fully and completely explained by this sort of explanation.

Leibniz has made this point very persuasively. "Let us suppose a book of the elements of geometry to have been eternal, one copy always having been taken down from an earlier one; it is evident that, even though a reason can be given for the present book out of a past one, nevertheless, out of any number of books, taken in order, going backwards, we shall never come upon a *full* reason; though we might well always wonder why there should have been such books from all time—why there were books at all, and why they were written in this manner. What is true of books is true also of the different states of the world; for what follows is in some way copied from what precedes...And so, however far you go back to earlier states, you will never find in those states *a full reason* why there should be any world rather than none, and why it should be such as it is."[5]

However, a moment's reflection will show that, if any type of explanation is merely preliminary and provisional, it is teleological explanation, since it presupposes a background that itself stands in need of explanation. If I account for the existence of the manufactured satellite by saying that it was made by some scientists for a certain purpose, then such an explanation can clarify the existence of the satellite only if I assume that there existed materials out of which the satellite was made, and scientists who made it for some purpose. It, therefore, does not matter what type of explanation we give, whether causal or teleological: either type, any type of explanation, will imply the existence of something by reference to which the *explicandum* can be explained. This in turn must be accounted for in the same way, and so on forever.

However, is not God a necessary being? Do we not escape the infinite regress as soon as we reach God? It is often maintained that, unlike ordinary intelligent beings, God is eternal and necessary; hence, His existence, unlike theirs, is not in need of explanation. For what is it that creates the vicious regress just mentioned? It is that, if we accept the principle of sufficient reason (that there must be an explanation for the existence of anything and everything the existence of which is not logically necessary, but merely contingent[6]), the existence of all the things referred to in any explanation requires itself to be explained. If, however, God is a logically necessary being, then His existence requires no explanation. Hence, the vicious regress comes to an end with God.

Now, it need not be denied that God is a necessary being in some sense of that expression. In one of these senses, I, for instance, am a necessary being: it is impossible that I should not exist, because it is self-refuting to say "I do not exist." The same is true of the English language and of the universe. It is self-refuting to say "There is no such thing as the English language" because this sentence is in the English language, or "There is no such thing as the universe" because whatever there is *is* the universe. It is impossible that these things should not in fact exist, since it is impossible that we should be mistaken in thinking that they exist. For what possible occurrence could even throw doubt on our being right on these matters, let alone show that we are wrong? I, the English language, and the universe are necessary beings, simply in the sense in which all is necessarily true that has been *proved* to be true. The occurrence of utterances such as "I exist," "The English language exists," and "The universe exists" is in itself sufficient proof of their truth. These remarks are, therefore, necessarily true, hence the things asserted to exist are necessary things.

However, this sort of necessity will not satisfy the principle of sufficient reason, because it is only hypothetical or consequential necessity.[7] *Given that* someone says "I exist," then it is logically impossible that *he* should not exist. Given the evidence we have, the English language and the universe most certainly do exist, but there is no necessity about the evidence. On the principle of sufficient reason, we must explain the existence of the evidence, for its existence is not logically necessary.

In other words, the only sense of "necessary being" capable of terminating the vicious regress is "logically necessary being," but it is no longer seriously in dispute that the notion of a logically necessary being is self-contradictory.[8] Whatever can be conceived of as existing can equally be conceived of as not existing.

However, even if there were such a thing as a logically necessary being, we could still not make out a case for the superiority of teleological over causal explanation. The existence of the universe cannot be explained in accordance with the familiar model of manufacture by a craftsman. For that model presupposes the existence of materials out of which the product is fashioned. God, on the other hand, must create the materials as well. Moreover, although we have a simple model of "creation out of nothing," for composers create tunes out of nothing, there is a great difference between creating *something to be sung,* and making the sounds that are a singing of it, or producing the piano on which to play it. Let us, however, waive all these objections and admit, for argument's sake, that creation out of nothing is conceivable. Surely, even so, no one can claim that it is the kind of explanation that yields the clearest and fullest understanding. Surely, to round off scientific explanations of the origin of the universe with creation out of nothing does not add anything to our understanding. There may be merit of some sort in this way of speaking, but whatever it is, it is not greater clarity or explanatory power.[9]

What then does all this amount to? Merely to the claim that scientific explanations are no worse than any other. All that has been shown is that all explanations suffer from the same defect: all involve a vicious infinite regress. In other words, no type of human explanation can help us to unravel the ultimate, unanswerable mystery. Christian ways of looking at things may not be able to render the world any more lucid than science can, but at least they do not pretend that there are no impenetrable mysteries. On the contrary, they point out untiringly that the claims of science to be able to elucidate everything are hollow. They remind us that science is not merely limited to the exploration of a tiny corner of the universe, but that, however far out probing instruments may eventually reach, we can never even approach the answers to the last questions: "Why is there a world at all rather than nothing?" and "Why is the world such as it is and not different?" Here, our finite human intellect bumps against its own boundary walls.

Is it true that scientific explanations involve an infinite vicious regress? Are scientific explanations really only provisional and incomplete? The crucial point will be this. Do *all* contingent truths call for explanation? Is the principle of sufficient reason sound? Can scientific explanations never come to a definite end? It will be seen that, with a clear grasp of the nature and purpose of explanation, we can answer these questions.[10]

Explaining something to someone is making him understand it. This involves bringing together in his mind two things, a model that is accepted as already simple and clear, and that which is to be explained, the *explicandum*, which is not so. Understanding the *explicandum* is seeing that it belongs to a range of things that could legitimately have been expected by anyone familiar with the model and with certain facts.

There are, however, two fundamentally different positions that a person may occupy relative to some *explicandum*. He may not be familiar with any model capable of leading him to expect the phenomenon to be explained. Most of us, for instance, are in that position in relation to the phenomena occurring in a good seance. With regard to other things, people will differ. Someone who can play chess, already understands chess and already has such a model. Someone who has never seen a game of chess has not. He sees the moves on the board, but he cannot understand, cannot follow, and cannot make sense of what is happening. Explaining the game to him is giving him an explanation, is making him understand. He can understand or follow chess moves only if he can see them as conforming to a model of a chess game. In order to acquire such a model, he will, of course, need to know the constitutive rules of chess, that is, the permissible moves, but that is not all. He must know that a normal game of chess is a competition (not all games are) between two people, each trying to win, and he must know what it is to win at chess: to maneuver the opponent's king into a position of checkmate. Finally, he must acquire some knowledge of what is and what is not conducive to winning: the tactical rules or canons of the game.

A person who has been given such an explanation and who has mastered it—which may take quite a long time—has now reached understanding, in the sense of the ability to follow each move. A person cannot, in that sense, understand merely one single move of chess and no other. If he does not understand any other moves, we must say that he has not yet mastered the explanation, and that he does not really understand the single move either. If he has mastered the explanation, then he understands all those moves that he can see as being in accordance with the model of the game inculcated in him during the explanation.

However, even though a person who has mastered such an explanation will understand many, perhaps most, moves of any game of chess he cares to watch, he will not necessarily understand them all, since some moves of a player may not be in accordance with his model of the game. White, let us say, at his fifteenth move, exposes his queen to capture by Black's knight. Though in accordance with the constitutive rules of the game, this move is, nevertheless, perplexing and calls for explanation, because it is not conducive to the achievement by White of what must be assumed to be his aim: to win the game. The queen is a much more valuable piece than the knight against which he is offering to exchange.

An onlooker who has mastered chess may fail to understand this move, be perplexed by it, and wish for an explanation. Of course, he may fail to be perplexed, for if he is a very inexperienced player, he may not see the disadvantageousness of the move. However, there is such a need whether anyone sees it or not. The move *calls* for explanation, because to anyone who knows the game, it must appear to be incompatible with the model that we have learned during the explanation of the game, and by reference to which we all explain and understand normal games.

However, the required explanation of White's fifteenth move is of a very different kind. What is needed now is not the acquisition of an explanatory model, but the removal of the real or apparent incompatibility between the player's move and the model of explanation he has already acquired. In such a case, the perplexity can be removed only on the assumption that the incompatibility between the model and the game is merely apparent. Since our model includes a presumed aim of both players, there are the following three possibilities:

1. White has made a mistake: he has overlooked the threat to his queen. In that case, the explanation is that White thought his move conducive to his end, but it was not.
2. Black has made a mistake: White set a trap for him. In that case, the explanation is that Black thought White's move was not conducive to White's end, but it was.
3. White is not pursuing the end that any chess player may be presumed to pursue: he is not trying to win his game. In that case, the explanation is that White has made a move that he

knows is not conducive to the end of winning his game because, let us say, he wishes to please Black who is his boss.

Let us now set out the differences and similarities between the two types of understanding involved in these two kinds of explanations. I shall call the first kind "model"—understanding and explaining, respectively, because both involve the use of a model by reference to which understanding and explaining are effected. The second kind I shall call "unvexing," because the need for this type of explanation and understanding arises only when there is a perplexity arising out of the incompatibility of the model and the facts to be explained.

The first point is that unvexing presupposes model-understanding, but not vice versa. A person can neither have nor fail to have unvexing-understanding of White's fifteenth move at chess, if he does not already have model-understanding of chess. Obviously, if I do not know how to play chess, I shall fail to have model-understanding of White's fifteenth move, but I can neither fail to have nor, of course, can I have unvexing-understanding of it, for I cannot be perplexed by it. I merely fail to have model-understanding of this move as, indeed, of any other move of chess. On the other hand, I may well have model-understanding of chess without having unvexing-understanding of every move. That is to say, I may well know how to play chess without understanding White's fifteenth move. A person cannot fail to have unvexing-understanding of the move unless he is vexed or perplexed by it; hence, he cannot even fail to have unvexing-understanding unless he already has model-understanding. It is not true that one either understands or fails to understand. On certain occasions, one neither understands nor fails to understand.

The second point is that there are certain things that cannot call for unvexing-explanations. No one can, for instance, call for an unvexing-explanation of White's first move, which is Pawn to King's Four, for no one can be perplexed or vexed by this move. Either a person knows how to play chess or he does not. If he does, then he must understand this move, for if he does not understand it, he has not yet mastered the game. If he does not know how to play chess, then he cannot yet have, or fail to have, unvexing-understanding, and he cannot, therefore, need an unvexing-explanation. Intellectual problems do not arise out of ignorance, but out of insufficient knowledge. An ignoramus is puzzled by very little. Once a student can see problems, he is already well into the subject.

The third point is that model-understanding implies being able, without further thought, to have model-understanding of a good many other things; unvexing-understanding does not. A person who knows chess and, therefore, has model-understanding of it must understand a good many chess moves, in fact all except those that call for unvexing-explanations. If he claims that he can understand White's first move, but no others, then he is either lying or deceiving himself or he really does not understand any move. On the other hand, a person who, after an unvexing-explanation, understands White's fifteenth move need not be able, without further explanation, to understand Black's or any other further move that calls for unvexing-explanation.

What is true of explaining deliberate and highly stylized human behavior, such as playing a game of chess, is also true of explaining natural phenomena, for what is characteristic of natural phenomena (that they recur in essentially the same way—that they are, so to speak, repeatable) is also true of chess games, as it is not of games of tennis or cricket. There is only one important difference: humans themselves have invented and laid down the rules of chess, as they have not invented or laid down the "rules or laws governing the behavior of things." This difference between chess and phenomena is important, for it adds another way to the three already mentioned,[11] in which a perplexity can be removed by an unvexing-explanation, namely, by abandoning the original explanatory model. This is, of course, not possible in the case of games of chess, because the model for chess is not a "construction" on the basis of the already existing phenomena of chess, but an invention.

The person who first thought up the model of chess could not have been mistaken. The person who first thought of a model explaining some phenomenon could have been mistaken.

Consider an example. We may think that the following phenomena belong together: the horizon seems to recede however far we walk towards it; we seem to be able to see further the higher the mountain we climb; the sun and moon seem every day to fall into the sea on one side but to come back from behind the mountains on the other side without being any the worse for it. We may explain these phenomena by two alternative models: (a) that the earth is a large disc; (b) that it is a large sphere. However, to a believer in the first theory there arises the following perplexity: how is it that, when we travel long enough towards the horizon in any one direction, we do eventually come back to our starting point without ever coming to the edge of the earth? We may at first attempt to "save" the model by saying that there is only an apparent contradiction. We may say either that the model does not require us to come to an edge, for it may be possible only to walk round and round on the flat surface, or we may say that the person must have walked over the edge without noticing it, or perhaps that the travelers are all lying. Alternatively, the fact that our model is "constructed" and not invented or laid down enables us to say, what we could not do in the case of chess, that the model is inadequate or unsuitable. We can choose another model that fits all the facts, for instance, that the earth is round. Of course, then we have to give an unvexing-explanation for why it looks flat, but we are able to do that.

We can now return to our original question, "Are scientific explanations true and full explanations or do they involve an infinite regress, leaving them forever incomplete?" Our distinction between model- and unvexing-explanations will help here. It is obvious that only those things that are perplexing *call for* and *can be given* unvexing-explanations. We have already seen that, in disposing of one perplexity, we do not necessarily raise another. On the contrary, unvexing-explanations truly and completely explain what they set out to explain, namely, how something is possible that, on our explanatory model, seemed to be impossible. There can, therefore, be no infinite regress here. Unvexing-explanations are real and complete explanations.

Can there be an infinite regress, then, in the case of model-explanations? Take the following example. European children are puzzled by the fact that their antipodean counterparts do not drop into empty space. This perplexity can be removed by substituting for their explanatory model another one. The European children imagine that throughout space there is an all-pervasive force operating in the same direction as the force that pulls them to the ground. We must, in our revised model, substitute for this force another acting everywhere in the direction of the center of the earth. Having thus removed their perplexity by giving them an adequate model, we can, however, go on to ask *why* there should be such a force as the force of gravity, why bodies should "naturally," in the absence of forces acting on them, behave in the way stated in Newton's laws, and we might be able to give such an explanation. We might, for instance, construct a model of space that would exhibit as derivable from it what in Newton's theory are "brute facts." Here we would have a case of the brute facts of one theory being explained within the framework of another, more general theory, and it is a sound methodological principle that we should continue to look for more and more general theories.

Note two points, however. The first is that we must distinguish, as we have seen, between *the possibility* and *the necessity* of giving an explanation. Particular occurrences can be explained by being exhibited as instances of regularities, and regularities can be explained by being exhibited as instances of more general regularities. Such explanations make things clearer. They organize the material before us. They introduce order where previously there was disorder, but absence of this sort of explanation (model-explanation) does not leave us with a puzzle or perplexity, an intellectual restlessness or cramp. The unexplained things are not unintelligible, incomprehensible, or irrational. Some things, on the other hand, call for, require and demand an explanation. As long as we are without such an explanation, we are perplexed, puzzled, and intellectually perturbed. We need an unvexing-explanation.

Now, it must be admitted that we may be able to construct a more general theory, from which, let us say, Newton's theory can be derived. This would further clarify the phenomena of motion and would be intellectually satisfying, but failure to do so would not leave us with an intellectual cramp. The facts stated in Newton's theory do not require, or stand in need of, unvexing-explanations. They could do so only if we already had another theory or model with which Newton's theory was incompatible. They could not do so by themselves, prior to the establishment of such another model.

The second point is that there is an objective limit to which such explanations tend, and beyond which they are pointless. There is a very good reason for wishing to explain a less general by a more general theory. Usually, such a unification goes hand in hand with greater precision in measuring the phenomena that both theories explain. Moreover, the more general theory, because of its greater generality, can explain a wider range of phenomena including not only phenomena already explained by some other theories, but also newly discovered phenomena, which the less general theory cannot explain. Now, the ideal limit to which such expansions of theories tend is an all-embracing theory that unifies all theories and explains all phenomena. Of course, such a limit can never be reached, since new phenomena are constantly discovered. Nevertheless, theories may be tending towards it. It will be remembered that the contention made against scientific theories was that there was no such limit because they involved an infinite regress. On that view, which I reject, there is no conceivable point at which scientific theories could be said to have explained the whole universe. On the view I am defending, there is such a limit, and it is the limit toward which scientific theories are actually tending. I claim that the nearer we come to this limit, the closer we are to a full and complete explanation of everything. For if we were to reach the limit, then though we could, of course, be left with a model that is itself unexplained and could be yet further explained by derivation from another model, there would be no need for, and no point in, such a further explanation. There would be no need for it, because any clearly defined model permitting us to expect the phenomena it is designed to explain offers full and complete explanations of these phenomena, however narrow the range. Also, although at lower levels of generality, there is a good reason for providing more general models, since they further simplify, systematize, and organize the phenomena, this, which is the only reason for building more general theories, no longer applies once we reach the ideal limit of an all-embracing explanation.

It might be said that there is another reason for using different models: that they might enable us to discover new phenomena. Theories are not only instruments of explanation, but also of discovery. With this I agree, but it is irrelevant to my point: that *the needs of explanation* do not require us to go on forever deriving one explanatory model from another.

It must be admitted, then, that in the case of model-explanations there is a regress, but it is neither vicious nor infinite. It is not vicious because, in order to explain a group of *explicanda*, a model-explanation *need* not itself be derived from another more general one. It gives a perfectly full and consistent explanation by itself. Also, the regress is not infinite, for there is a natural limit, an all-embracing model, that can explain all phenomena, beyond which it would be pointless to derive model-explanations from yet others.

What about our most serious question, "Why is there anything at all?" Sometimes, when we think about how one thing has developed out of another and that one out of a third, and so on back throughout all time, we are driven to ask the same question about the universe as a whole. We want to add up all things and refer to them by the name, "the world," and we want to know why the world exists and why there is not nothing instead. In such moments, the world seems to us a kind of bubble floating on an ocean of nothingness. Why should such flotsam be adrift in empty space? Surely, its emergence from the hyaline billows of nothingness is more mysterious even than Aphrodite's emergence from the sea. Wittgenstein expressed in these words the mystification we all feel: "Not *how* the world is, is the mystical, but *that* it is. The contemplation of the world *sub specie aeterni* is the contemplation of it as a limited whole. The feeling of the world as a limited whole is the mystical feeling."[12]

Professor J. J. C. Smart expresses his own mystification in these moving words:

> That anything should exist at all does seem to me a matter for the deepest awe. But whether other people feel this sort of awe, and whether they or I ought to is another question. I think we ought to. If so, the question arises: If "Why should anything exist at all?" cannot be interpreted after the manner of the cosmological argument, that is, as an absurd request for the non-sensical postulation of a logically necessary being, what sort of question is it? What sort of question is this question "Why should anything exist at all?" All I can say is that I do not yet know.[13]

It is undeniable that the magnitude and perhaps the very existence of the universe are awe-inspiring. It is probably true that it gives many people "the mystical feeling." It is also undeniable that our awe, our mystical feeling, aroused by contemplating the vastness of the world, is justified, in the same sense in which our fear is justified when we realize we are in danger. There is no more appropriate object for our awe or for the mystical feeling than the magnitude and perhaps the existence of the universe, just as there is no more appropriate object for our fear than a situation of personal peril. However, it does not follow from this that it is a good thing to cultivate, or indulge in, awe or mystical feelings, any more than it is necessarily a good thing to cultivate, or indulge in, fear in the presence of danger.

In any case, whether or not we ought to have or are justified in having a mystical feeling or a feeling of awe when contemplating the universe, having such a feeling is not the same as asking a meaningful question, although having it may well *incline* us to utter certain forms of words. Our question, "Why is there anything at all?" may be no more than the expression of our feeling of awe or mystification, and not a meaningful question at all. Just as the feeling of fear may naturally but illegitimately give rise to the question "What sin have I committed," so the feeling of awe or mystification may naturally but illegitimately lead to the question, "Why is there anything at all?" What we have to discover, then, is whether this question makes sense or is meaningless.

Yes, of course, it will be said, it makes perfectly good sense. There is an undeniable fact and it calls for explanation. The fact is that the universe exists. In the light of our experience, there can be no possible doubt that something or other exists, and the claim that the universe exists commits us to no more than that. Surely this calls for explanation, because the universe must have originated somehow. Everything has an origin and the universe is no exception. Since the universe is the totality of things, it must have originated out of nothing. If it had originated out of something, even something as small as one single hydrogen atom, what has so originated could not be the whole universe, but only the universe minus the atom. Then the atom itself would call for explanation, for it too must have had an origin, and it must be *an origin out of nothing.* Also, how can anything originate out of nothing? Surely that calls for explanation.

However, let us be quite clear what is to be explained. There are two facts here, not one. The first is that the universe exists, which is undeniable. The second is that the universe must have originated out of nothing, and that is not undeniable. It is true that, *if it has originated at all,* then it must have originated out of nothing, or else it is not the universe that has originated. However, need it have originated? Could it not have existed forever?[14] It might be argued that nothing exists forever, that everything has originated out of something else. That may well be true, but it is perfectly compatible with the fact that the universe is everlasting. We may well be able to trace the origin of any thing to the time when, by some transformation, it has developed out of some other thing, and yet it may be the case that no thing has its origin in nothing, and the universe has existed forever. For even if every *thing* has a beginning and an end, the total of mass and energy may well remain constant.

Moreover, the hypothesis that the universe originated out of nothing is, empirically speaking, completely empty. Suppose, for argument's sake, that the annihilation of an object without remainder is conceivable. It would still not be possible for any hypothetical observer to ascertain whether space was empty or not. Let us suppose that, *within the range of observation of our observer,* one object after another is annihilated without remainder and that only one is left. Our observer could not then tell whether in remote parts of the universe, beyond his range of observation, objects are coming into being or passing out of existence. What, moreover, are we to say of the observer himself? Is he to count for nothing? Must we not postulate him away as well, if the universe is to have arisen out of nothing?

Let us, however, ignore all these difficulties and assume that the universe really has originated out of nothing. Even that does not prove that the universe has not existed forever. If the universe can conceivably develop out of nothing, then it can conceivably vanish without remainder, and it can arise out of nothing again and subside into nothingness once more, and so on *ad infinitum.* Of course, "again" and "once more" are not quite the right words. The concept of time hardly applies to such universes. It does not make sense to ask whether one of them is earlier or later than, or perhaps simultaneous with, the other because we cannot ask whether they occupy the same or different spaces. Being separated from one another by "nothing," they are not separated from one another by "anything." We cannot, therefore, make any statements about their mutual spatio-temporal relations. It is impossible to distinguish between one long continuous universe and two universes separated by nothing. How, for instance, can we tell whether the universe including ourselves is not frequently annihilated and "again" reconstituted just as it was?

Let us now waive these difficulties as well. Let us suppose for a moment that we understand what is meant by saying that the universe originated out of nothing and that this has happened only once. Let us accept this as a fact. Does this fact call for explanation?

It does not call for an unvexing-explanation. That would be called for only if there were a perplexity owing to the incompatibility of an accepted model with some fact. In our case, the fact to be explained is the origination of the universe out of nothing; hence, there could not be such a perplexity, for we need not employ a model incompatible with this. If we had a model incompatible with our "fact," then that would be the wrong model and we would simply have to substitute another for it. The model we employ to explain the origin of the universe out of nothing could not be based on the similar origins of other things for, of course, there is nothing else with a similar origin.

All the same, it seems very surprising that something should have come out of nothing. It is contrary to the principle that every thing has an origin, that is, has developed out of something else. It must be admitted that there is this incompatibility. However, it does not arise because a well-established model does not square with an undeniable fact; it arises because a well-established model does not square with *an assumption* of which it is hard even to make sense and for which there is no evidence whatsoever. In fact, the only reason we have for making this assumption is a simple logical howler: that because every thing has an origin, the universe must have an origin, too, except that, being the universe, it must have originated out of nothing. This is a howler, because it conceives of the universe as a big thing, whereas, in fact, it is the totality of things, that is, not a thing. That every thing has an origin does not entail that the totality of things has an origin. On the contrary, it strongly suggests that it has not. For to say that every thing has an origin implies that any given thing must have developed out of something else, which in turn, being a thing, must have developed out of something else, and so forth. If we assume that every thing has an origin, we need not, indeed it is hard to see how we can, assume that the totality of things has an origin as well. There is, therefore, no perplexity, because we need not and should not assume that the universe has originated out of nothing.

If, however, in spite of all that has been said just now, someone still wishes to assume, contrary to all reason, that the universe has originated out of nothing, there would still be no perplexity, for then he would simply have to give up the principle that is incompatible with this assumption, namely, that no thing can originate out of nothing. After all, this principle *could* allow for exceptions. We have no proof that it does not. Again, there is no perplexity, because there is no incompatibility between our assumption and an inescapable principle.

However, it might be asked, do we not need a model-explanation of our supposed fact? The answer is No. We do not need such an explanation, for there could not possibly be a model for this origin other than this origin itself. We cannot say that origination out of nothing is like birth, or emergence, or evolution, or anything else we know, for it is not like anything we know. In all these cases, there is *something* out of which the new thing has originated.

To sum up, the question, "Why is there anything at all?" looks like a perfectly sensible question modeled on "Why does *this* exist?" or "How has *this* originated?" It looks like a question about the origin of a thing. However, it is not such a question, for the universe is not a thing, but the totality of things. There is, therefore, no reason to assume that the universe has an origin. The very assumption that it has is fraught with contradictions and absurdities. If, nevertheless, it were true that the universe originated out of nothing, then this would not call either for an unvexing- or a model-explanation. It would not call for the latter, because there could be no model of it taken from another part of our experience, since there is nothing analogous in our experience to origination out of nothing. It would not call for the former, because there can be no perplexity because of the incompatibility of a well-established model and an undeniable fact, since there is no undeniable fact and no well-established model. If, on the other hand, as is more probable, the universe has not originated at all, but is eternal, then the question why or how it has originated simply does not arise. There can then be no question about why anything at all exists, for it could not mean how or why the universe originated, since *ex hypothesi* it has no origin. What else could it mean?

Lastly, we must bear in mind that the hypothesis that the universe was made by God out of nothing only brings us back to the question who made God or how God originated, and if we do not find it repugnant to say that God is eternal, we cannot find it repugnant to say that the universe is eternal. The only difference is that we know for certain that the universe exists, whereas we have the greatest difficulty in even making sense of the claim that God exists.

To sum up, according to the argument examined, we must reject the scientific world picture because it is the outcome of scientific types of explanations that do not really and fully explain the world around us, but only tell us *how* things have come about, not *why,* and can give no answer to the ultimate question, why there is anything at all rather than nothing. Against this, I have argued that scientific explanations are real and full, just like the explanations of everyday life and of the traditional religions. They differ from the latter only in that they are more precise and more easily disprovable by the observation of facts.

My main points dealt with the question why scientific explanations were thought to be merely provisional and partial. The first main reason is the misunderstanding of the difference between teleological and causal explanations. It is first, and rightly, maintained that teleological explanations are answers to "Why?"questions, whereas causal explanations are answers to "How?"questions. It is further, and wrongly, maintained that, in order to obtain real and full explanations of anything, one must answer both "Why?" and "How?" questions. In other words, it is thought that all matters can and must be explained by both teleological and causal types of explanations. Causal explanations, it is believed, are merely provisional and partial, waiting to be completed by teleological explanations. Until a teleological explanation has been given, so the story goes, we have not *really* understood the *explicandum.* However, I have shown that both types are equally real and full explanations. The difference between them is merely that they are appropriate to different types of *explicanda.*

It should, moreover, be borne in mind that teleological explanations are not, in any sense, unscientific. They are rightly rejected in the natural sciences, not, however, because they are unscientific, but because no intelligences or purposes are found to be involved there. On the other hand, teleological explanations are very much in place in psychology, for we find intelligence and purpose involved in a good deal of human behavior. It is not only not unscientific to give teleological explanations of deliberate human behavior, but it would be quite unscientific to exclude them.

The second reason why scientific explanations are thought to be merely provisional and partial is that they are believed to involve a vicious infinite regress. Two misconceptions have led to this important error. The first is the general misunderstanding of the nature of explanation, and in particular, the failure to distinguish between the two types that I have called model- and unvexing-explanations. If one does not draw this distinction, it is natural to conclude that scientific explanations lead to a vicious infinite regress. For although it is true of those perplexing matters, which are elucidated by unvexing-explanations, that they are incomprehensible and cry out for explanation, it is not true that, after an unvexing-explanation has been given, this itself is again capable, let alone in need of, a yet further explanation of the same kind. Conversely, although it is true that model-explanations of regularities can themselves be further explained by more general model-explanations, it is not true that, in the absence of such more general explanations, the less general are incomplete, hang in the air, so to speak, leaving the *explicandum* incomprehensible and crying out for explanation. The distinction between the two types of explanations shows us that an *explicandum* is either perplexing and incomprehensible, in which case an explanation of it is *necessary* for clarification and, when given, *complete,* or it is a regularity capable of being subsumed under a model, in which case a further explanation *is possible* and often profitable, but *not necessary* for clarification.

The second misconception responsible for the belief in a vicious infinite regress is the misrepresentation of scientific explanation *as essentially causal.* It has generally been held that, in a scientific explanation, the *explicandum* is the effect of some event, the cause, temporally prior to the *explicandum*. Combined with the principle of sufficient reason (the principle that anything is in need of explanation that might conceivably have been different from what it is), this error generates the nightmare of determinism. Since any event might have been different from what it was, acceptance of this principle has the consequence that *every* event must have a reason or explanation. However, if the reason is itself an event *prior in time,* then every reason must have a reason preceding it, and so the infinite regress of explanation is necessarily tied to the time scale stretching infinitely back into the endless past. It is, however, obvious from our account that science is not primarily concerned with the forging of such causal chains. The primary object of the natural sciences is not historical at all. Natural science claims to reveal, not the beginnings of things, but their underlying reality. It does not dig up the past; it digs down into the structure of things existing here and now. Some scientists do allow themselves to speculate, and rather precariously at that, about origins, but their hard work is done on the structure of what exists now. In particular, those explanations that are themselves further explained are not explanations linking event to event in a gapless chain reaching back to creation day, but generalizations of theories tending toward a unified theory.

The Purpose of Human Existence

Our conclusion in the previous section has been that science is, in principle, able to give complete and real explanations of every occurrence and thing in the universe. This has two important corollaries: (i) Acceptance of the scientific world picture cannot be *one's reason* for the belief that the universe is unintelligible and, therefore, meaningless, though coming to accept it, after having been taught the Christian world picture, may well be, in the case of many individuals, *the only or the main cause* of their belief that the universe and human existence are meaningless. (ii) It is not in accordance with reason to reject this pessimistic belief on the grounds that scientific explanations are only provisional and incomplete, and must be supplemented by religious ones.

In fact, it might be argued that the more clearly we understand the explanations given by science, the more we are driven to the conclusion that human life has no purpose and, therefore, no meaning. The science of astronomy teaches us that our earth was not specially created about 6000 years ago, but evolved out of hot nebulae that previously had whirled aimlessly through space for countless ages. As they cooled, the sun and the planets formed. On one of these planets at a certain time the circumstances were propitious and life developed. However, conditions will not remain favorable to life. When our solar system grows old, the sun will cool, our planet will be covered with ice, and all living creatures will eventually perish. Another theory has it that the sun will explode and that the heat generated will be so great that all organic life on earth will be destroyed. That is the comparatively short history and prospect of life on earth. Altogether it amounts to very little when compared with the endless history of the inanimate universe.

Biology teaches us that the human species was not specially created but is merely, in a long chain of evolutionary changes of forms of life, the last link, made in the likeness, not of God, but of nothing so much as an ape. The rest of the universe, whether animate or inanimate, instead of serving the ends of humans, is at best indifferent and at worst savagely hostile. Evolution of whose operation the emergence of humans is the result is a ceaseless battle among members of different species, one species being gobbled up by another, only the fittest surviving. Far from being the gentlest and most highly moral, humans are simply the creatures best fitted to survive, the most efficient, if not the most rapacious and insatiable killers. Also, in this unplanned, fortuitous, monstrous, savage world, humans are madly trying to snatch a few brief moments of joy in the short intervals during which they are free from pain, sickness, persecution, war, or famine until, finally, their lives are snuffed out in death. Science has helped us to know and understand this world, but what purpose or meaning can it find in it?

Complaints such as these do not mean quite the same thing to everybody, but one thing, I think, they mean to most people is: science shows life to be meaningless, because life is without purpose. The medieval world picture provided life with a purpose; hence, medieval Christians could believe that life had a meaning. The scientific account of the world takes away life's purpose and with it its meaning.

There are, however, two quite different senses of "purpose." Which one is meant? Has science deprived human life of purpose in both senses? If not, is it a harmless sense, in which human existence has been robbed of purpose? Could human existence still have meaning if it did not have a purpose in that sense?

What are the two senses? In the first and basic sense, purpose is normally attributed only to persons or their behavior, as in "Did you have a purpose in leaving the ignition on?" In the second sense, purpose is normally attributed only to things, as in "What is the purpose of that gadget you installed in the workshop?" The two uses are intimately connected. We cannot attribute a purpose to a thing without implying that someone did something, in the doing of which he had some purpose, namely, to bring about the thing with the purpose. Of course, *his* purpose is not identical with *its* purpose. In hiring laborers and engineers and buying materials and a site for a factory and the like, the entrepreneur's purpose, let us say, is to manufacture cars, but the purpose of cars is to serve as a means of transportation.

There are many things that humans may do, such as buying and selling, hiring laborers, ploughing, felling trees, and the like, which it is foolish, pointless, silly, and perhaps crazy to do if one has no purpose in doing them. A person who does these things without a purpose is engaging in inane, futile pursuits. Lives crammed full with such activities, devoid of purpose, are pointless, futile, and worthless. Such lives may indeed be dismissed as meaningless, but it should also be perfectly clear that acceptance of the scientific world picture does not force us to regard our lives as being without a purpose in this sense. Science has not only not robbed us of any purpose that we had before, but it has furnished us with enormously greater power to achieve these purposes. Instead of praying for rain or a good harvest or offspring, we now use ice pellets, artificial manure, or artificial insemination.

By contrast, having or not having a purpose, in the other sense, is value neutral. We do not think more or less highly of a thing for having or not having a purpose. "Having a purpose," in this sense, confers no kudos; "being purposeless" carries no stigma. A row of trees growing near a farm may or may not have a purpose: it may or may not be a windbreak, may or may not have been planted or deliberately left standing there in order to prevent the wind from sweeping across the fields. We do not, in any way, disparage the trees if we say they have no purpose, but have just grown that way. They are as beautiful, made of as good wood, and as valuable as they would be if they had a purpose. Also, of course, they break the wind just as well. The same is true of living creatures. We do not disparage a dog when we say that it has no purpose, is not a sheep dog or a watch dog or a rabbiting dog, but just a dog that hangs around the house and is fed by us.

Humans are in a different category, however. To attribute to a human being a purpose in that sense is not neutral, let alone complimentary: it is offensive. It is degrading for a person to be regarded as merely serving a purpose. If, at a garden party, I ask a man in livery, "What is your purpose?" I am insulting him. I might as well have asked, "What are you *for?*" Such questions reduce him to the level of a gadget, a domestic animal, or perhaps a slave. I imply that *we* allot to *him* the tasks, the goals, and the aims that he is to pursue—that *his* wishes, desires, aspirations, and purposes are to count for little or nothing. We are treating him, in Kant's phrase, merely as a means to our ends, not as an end in himself.

The Christian and the scientific world pictures do indeed differ fundamentally on this point. The latter robs humans of a purpose in this sense. It sees them as beings with no purpose allotted to them by anyone but themselves. It robs them of any goal, purpose, or destiny appointed for them by any outside agency. The Christian world picture, on the other hand, sees humans as creatures, divine artifacts, something halfway between robots (manufactured) and animals (alive), *homunculi,* or perhaps Frankenstein's monsters, made in God's laboratory, with a purpose or task assigned by their Maker.

However, lack of purpose in this sense does not in any way detract from the meaningfulness of life. I suspect that many who reject the scientific outlook because it involves the loss of purpose of life, and, therefore, meaning, are guilty of a confusion between the two senses of "purpose" just distinguished. They confusedly think that, if the scientific world picture is true, then their lives must be futile because that picture implies that humans have no purpose given from without. However, this is muddled thinking, for, as has already been shown, pointlessness is implied only by purposelessness in the other sense, which is not at all implied by the scientific picture of the world. These people mistakenly conclude that there can be no purpose *in* life because there is no purpose *of* life—that *people* cannot themselves adopt and achieve purposes because *humans,* unlike robots or watch dogs, are not creatures with a purpose.[15]

However, not all people taking this view are guilty of the above confusion. Some really hanker after a purpose of life in this sense. To some people, the greatest attraction of the medieval world picture is the belief in an omnipotent, omniscient, and all-good Father, the view of themselves as His children who worship Him, of their proper attitude to what befalls them as submission, humility, resignation in His will, and what is often described as the "creaturely feeling."[16] All these are attitudes and feelings appropriate to a being that stands to another in the same sort of relation, though of course on a higher plane, in which a helpless child stands to his progenitor. Many regard the scientific picture of the world as cold, unsympathetic, unhomely, and frightening, because it does not provide for any appropriate object of this creaturely attitude. There is nothing and no one in the world, as science depicts it, in which we can have faith or trust, on whose guidance we can rely, to whom we can turn for consolation, or whom we can worship or submit to—except other human beings. This may be felt as a keen disappointment, because it shows that the meaning of life cannot lie in submission to His will, in acceptance of whatever may come, and in worship, but it does not imply that life can have *no* meaning. It merely implies that it must have a different meaning from that which it was thought to have. Just as it is a great shock for a child to find that he must stand on his own feet, that his father and mother no longer provide for him, so a person who has lost his faith in God must reconcile himself to the idea that he has to stand on his own feet, alone in the world except for whatever friends he may succeed in making.

Is not this to miss the point of the Christian teaching? Surely, Christianity can tell us the meaning of life because it tells us the grand and noble end for which God has created the universe and humans. No human life, however pointless it may seem, is meaningless, because in being part of God's plan, every life is assured of significance.

This point is well taken. It brings to light a distinction of some importance: we call a person's life meaningful not only if it is worthwhile, but also if he has helped in the realization of some plan or purpose transcending his own concerns. A person who knows he must soon die a painful death can give significance to the remainder of his doomed life by, say, allowing certain experiments to be performed on him that will be useful in the fight against cancer. In a similar way, only on a much more elevated plane, every person, however humble or plagued by suffering, is guaranteed significance by the knowledge that he is participating in God's purpose.

What, then, on the Christian view, is the grand and noble end for which God has created the world and humans in it? We can immediately dismiss that still popular opinion that the smallness of our intellect prevents us from stating meaningfully God's design in all its imposing grandeur.[17] This view cannot possibly be a satisfactory answer to our question about the purpose of life. It is, rather, a confession of the impossibility of giving one. If anyone thinks that this "answer" can remove the sting from the impression of meaninglessness and insignificance in our lives, he cannot have been stung very hard.

If, then, we turn to those who are willing to state God's purpose in so many words, we encounter two insuperable difficulties. The first is to find a purpose grand and noble enough to explain and justify the great amount of undeserved suffering in this world. We are inevitably filled by a sense of bathos when we read statements such as this: "...history is the scene of a divine purpose, in which the whole history is included, and Jesus of Nazareth is the center of that purpose, both as revelation and as achievement, as the fulfillment of all that was past, and the promise of all that was to come. ...If God is God, and if He made all these things, why did He do it?...God created a universe, bounded by the categories of time, space, matter, and causality, because He desired to enjoy for ever the society of a fellowship of finite and redeemed spirits which have made to His love the response of free and voluntary love and service."[18] Surely this cannot be right? Could a God be called omniscient, omnipotent, and all-good who, for the sake of satisfying his desire to be loved and served, imposes (or has to impose) on his creatures the amount of undeserved suffering we find in the world?

There is, however, a much more serious difficulty still: God's purpose in making the universe must be stated in terms of a dramatic story, many of whose key incidents symbolize religious conceptions and practices that we no longer find morally acceptable: the imposition of a taboo on the fruits of a certain tree, the sin and guilt incurred by Adam and Eve by violating the taboo, the wrath of God,[19] the curse of Adam and Eve and all their progeny, the expulsion from Paradise, the Atonement by Christ's bloody sacrifice on the cross that makes available by way of the sacraments God's Grace by which alone humans can be saved (thereby, incidentally, establishing the valuable power of priests to forgive sins, and thus, alone make possible a person's entry to heaven), Judgment Day on which the sheep are separated from the goats and the latter condemned to eternal torment in hell-fire.

Obviously, it is much more difficult to formulate a purpose for creating the universe and humans that will justify the enormous amount of undeserved suffering that we find around us, if that story has to be fitted in as well. For now we have to explain not only why an omnipotent, omniscient, and all-good God should create such a universe and such a person, but also why, foreseeing every move of the feeble, weak-willed, ignorant, and covetous creature to be created, He should, nevertheless, have created him and, having done so, should be incensed and outraged by people's sin, and why He should deem it necessary to sacrifice His own son on the cross to atone for this sin that was, after all, only a disobedience of one of his commands, and why this atonement and consequent redemption could not have been followed by a return to Paradise—particularly of those innocent children who had not yet sinned—and why, on Judgment Day, this merciful God should condemn some to eternal torment.[20] It is not surprising that, in the face of these and other difficulties, we find, again and again, a return to the first view: that God's purpose cannot meaningfully be stated.

It will perhaps be objected that no Christian today believes in the dramatic history of the world as I have presented it, but this is not so. It is the official doctrine of the Roman Catholic, the Greek Orthodox, and a large section of the Anglican Church,[21] nor does Protestantism substantially alter this picture. In fact, by insisting on "Justification by Faith Alone" and by rejecting the ritualistic, magical character of the medieval Catholic interpretation of certain elements in the Christian religion, such as indulgences, the sacraments, and prayer, while insisting on the necessity of grace, Protestantism undermined the moral element in medieval Christianity expressed in the Catholics' emphasis on personal merit.[22] Protestantism, by harking back to St. Augustine, who clearly realized the incompatibility of grace and personal merit,[23] opened the way for Calvin's doctrine of Predestination (the intellectual parent of that form of rigid determinism that is usually blamed on science) and Salvation or Condemnation from all eternity.[24] Since Roman Catholics, Lutherans, Calvinists, Presbyterians, and Baptists officially subscribe to the views just outlined, one can justifiably claim that the overwhelming majority of professing Christians hold or ought to hold them.

It might still be objected that the best and most modern views are wholly different. I have not the necessary knowledge to pronounce on the accuracy of this claim. It may well be true that the best and most modern views are such as Professor Braithwaite's, who maintains that Chistianity is, roughly speaking, "morality plus stories," where the stories are intended merely to make the strict moral teaching both more easily understandable and more palatable,[25] or it may be that one or the other of the modern views on the nature and importance of the dramatic story told in the sacred Scriptures is the best. My reply is that, even if it is true, it does not prove what I wish to disprove, that one can extract a sensible answer to our question, "What is the meaning of life?" from the kind of story subscribed to by the overwhelming majority of Christians, who would, moreover, reject any such modernist interpretation at least as indignantly as the scientific account. Moreover, though such views can perhaps avoid some of the worst absurdities of the traditional story, they are hardly in a much better position to state the purpose for which God has created the universe and humans in it, because they cannot overcome the difficulty of finding a purpose grand and noble enough to justify the enormous amount of undeserved suffering in the world.

Let us, however, for argument's sake, waive all these objections. There remains one fundamental hurdle that no form of Christianity can overcome: the fact that it demands of humans a morally repugnant attitude towards the universe. It is now very widely held[26] that the basic element of the Christian religion is an attitude of worship toward a being supremely worthy of being worshipped, and that it is religious feelings and experiences that apprise their owner of such a being and that inspire in him the knowledge or the feeling of complete dependence, awe, worship, mystery, and self-abasement. There is, in other words, a bipolarity (the famous "I–Thou relationship") in which the object, "the wholly other," is exalted whereas the subject is abased to the limit. Rudolf Otto has called this the "creature-feeling,"[27] and he quotes as an expression of it Abraham's words when venturing to plead for the men of Sodom: "Behold now, I have taken upon me to speak unto the Lord, which am but dust and ashes." (Gen. XVIII.27). Christianity thus demands of people an attitude inconsistent with one of the presuppositions of morality: that humans are not wholly dependent on something else, that they have free will, that they are, in principle, capable of responsibility. We have seen that the concept of grace is the Christian attempt to reconcile the claim of total dependence and the claim of individual responsibility (partial independence), and it is obvious that such attempts must fail. We may dismiss certain doctrines, such as the doctrine of original sin, the doctrine of eternal hell-fire, or the doctrine that there can be no salvation outside the Church as extravagant and peripheral, but we cannot reject the doctrine of total dependence without rejecting the characteristically Christian attitude as such.

The Meaning of Life

Perhaps some of you will have felt that I have been shirking the real problem. To many people, the crux of the matter seems as follows: How can there be any meaning in our lives if they end in death? What meaning can there be in them that our inevitable deaths do not destroy? How can our existences be meaningful if there is no afterlife in which perfect justice is meted out? How can life have any meaning if all it holds out to us are a few miserable earthly pleasures, and even these to be enjoyed only rarely and for such a piteously short time?

I believe this is the point that exercises most people most deeply. Kirilov, in Dostoyevsky's novel *The Possessed,* claims, just before committing suicide, that as soon as we realize that there is no God, we cannot live any longer—we must put an end to our lives. One of the reasons that he gives is that, when we discover that there is no paradise, we have nothing to live for.

> ...there was a day on earth, and in the middle of the earth were three crosses. One on the cross had such faith that He said to another, "To-day thou shalt be with me in paradise." The day came to an end, both died, and they went, but they found neither paradise nor resurrection. The saying did not come true. Listen: that man was the highest of all on earth. ...There has never been any one like Him before or since, and never will be. ...And if that is so, if the laws of Nature did not spare even *Him,* and made even Him live in the midst of lies and die for a lie, then the whole planet is a lie and is based on a lie and a stupid mockery. So the very laws of the planet are a lie and a farce of the devil. What, then, is there to live for?[28]

Tolstoy, too, was nearly driven to suicide when he came to doubt the existence of God and an afterlife,[29] and this is true of many.

What, then, is it that inclines us to think that, if life is to have a meaning, there would be an afterlife? It is this. The Christian world view contains the following three propositions. The first is that, since the Fall, God's curse of Adam and Eve, and the expulsion from Paradise, life on earth for humankind has not been worthwhile, but a vale of tears, and one long chain of misery, suffering, unhappiness, and injustice. The second is that a perfect afterlife is awaiting us after the death of the body. The third is that we can enter this perfect life only on certain conditions, among which is also the condition of enduring our earthly existence to its bitter end. In this way, our earthly existence, which, in itself, would not (at least for many people if not all) be worth living, acquires meaning and significance; only if we endure it can we gain admission to the realm of the blessed.

It might be doubted whether this view is still held today. However, there can be no doubt that even today we all imbibe a good deal of this view with our earliest education. In sermons, the contrast between the perfect life of the blessed and our life of sorrow and drudgery is frequently driven home, and we hear again and again that Christianity has a message of hope and consolation for all those "who are weary and heavy laden."[30]

It is not surprising then that, when the implications of the scientific world picture begin to sink in, when we come to have doubts about the existence of God and another life, we are bitterly disappointed. For if there is no afterlife, then all we are left is our earthly life, which we have come to regard as a necessary evil, the painful fee of admission to the land of eternal bliss, but if there is no eternal bliss to come and if this hell on earth is all, why hang on till the horrible end?

Our disappointment, therefore, arises out of these two propositions, that the earthly life is not worth living, and that there is another perfect life of eternal happiness and joy that we may enter upon if we satisfy certain conditions. We can regard our lives as meaningful, if we believe both. We cannot regard them as meaningful if we believe merely the first and not the second. It seems to me inevitable that people who are taught something of the history of science will have serious doubts about the second. If they cannot overcome these, as many will be unable to do, then they must either accept the sad view that their lives are meaningless or they must abandon the first proposition: that this earthly life is not worth living. They must find the meaning of their lives in this earthly existence, but is this possible?

A moment's examination will show us that the Christian evaluation of our earthly lives as worthless, which we accept in our moments of pessimism and dissatisfaction, is not one that we normally accept. Consider only the question of murder and suicide. On the Christian view, other things being equal, the most kindly thing to do would be for every one of us to kill as many of our friends and dear ones as still have the misfortune to be alive, and then to commit suicide without delay, for every moment spent in this life is wasted. On the Christian view, God has not made it that easy for us. He has forbidden us to hasten others or ourselves into the next life. Our bodies are his private property and must be allowed to wear themselves out in the way decided by Him, however painful and horrible that may be. We are, as it were, driving a burning car. There is only one way out, to jump clear and let it hurtle to destruction. However, the owner of the car has forbidden it on pain of eternal tortures worse than burning, and so we do better to burn to death inside.

On this view, murder is a less serious wrong than suicide. For murder can always be confessed and repented and, therefore, forgiven; suicide cannot—unless we allow the ingenious way out chosen by the heroine of Graham Greene's play "The Living Room," who swallows a slow but deadly poison and, while awaiting its taking effect, repents having taken it. Murder, on the other hand, is not so serious because, in the first place, it need not rob the victim of anything but the last lap of his march in the vale of tears, and, in the second place, it can always be forgiven. Hamlet, it will be remembered, refrains from killing his uncle during the latter's prayers because, as a true Christian, he believes that killing his uncle at that point, when the latter had purified his soul by repentance, would merely be doing him a good turn, for murder at such a time would simply dispatch him to undeserved and everlasting happiness.

These views strike us as odd, to say the least. They are the logical consequence of the official medieval evaluation of this our earthly existence. If this life is not worth living, then taking it is not robbing the person concerned of much. The only thing wrong with it is the damage to God's property, which is the same both in the case of murder and suicide. We do not take this view at all. Our view, on the contrary, is that murder is the most serious wrong because it consists of taking away from someone else against his will his most precious possession, his life. For this reason, when a person suffering from an incurable disease asks to be killed, the mercy killing of such a person is regarded as a much less serious crime than murder because, in such a case, the killer is not robbing the other of a good against his will. Suicide is not regarded as a real crime at all, for we take the view that a person can do with his own possessions what he likes.

However, from the fact that these are our normal opinions, we can infer nothing about their truth. After all, we could easily be mistaken. Whether life is or is not worthwhile, is a value judgment. Perhaps all this is merely a matter of opinion or taste. Perhaps no objective answer can be given. Fortunately, we need not enter deeply into these difficult and controversial questions. It is quite easy to show that the medieval evaluation of earthly life is based on a misguided procedure.

Let us remind ourselves briefly of how we arrive at our value judgments. When we determine the merits of students, meals, tennis players, bulls, or bathing belles, we do so on the basis of some criteria and some standard or norm. Criteria and standards notoriously vary from field to field and even from case to case, but that does not mean that we have no idea about what the appropriate criteria or standards to use are. It would not be fitting to apply the criteria for judging bulls to the judgment of students or bathing belles. They score on quite different points, and even where the same criteria are appropriate, as in the judgment of students enrolled in different schools and universities, the standards will vary from one institution to another. Pupils who would only just pass in one would perhaps obtain honors in another. The higher the standard applied, the lower the marks, that is, the merit conceded to the candidate.

The same procedure is applicable also in the evaluation of a life. We examine it on the basis of certain criteria and standards. The medieval Christian view uses the criteria of the ordinary human: a life is judged by what the person concerned can get out of it: the balance of happiness over unhappiness, pleasure over pain, bliss over suffering. Our earthly lives are judged not worthwhile because they contain much unhappiness, pain, and suffering, and little happiness, pleasure, and bliss. The next life is judged worthwhile because it provides eternal bliss and no suffering.

Armed with these criteria, we can compare the life of this person and that, and judge which is more worthwhile and which has a greater balance of bliss over suffering, but criteria alone enable us merely to make comparative judgments of value, not absolute ones. We can say which is more and which is less worthwhile, but we cannot say which is worthwhile and which is not. In order to determine the latter, we must introduce a standard, but what standard ought we to choose?

Ordinarily, the standard we employ is the average of the kind. We call a human and a tree tall if they are well above the average of their kind. We do not say that Jones is a short man because he is shorter than a tree. We do not judge a boy a bad student because his answer to a question in the Leaving Examination is much worse than that given in reply to the same question by a young man sitting for his finals for the Bachelor's degree.

The same principles must apply to judging lives. When we ask whether a given life was or was not worthwhile, then we must take into consideration the range of worthwhileness that ordinary lives normally cover. Our end poles of the scale must be the best possible and the worst possible life that one finds. A good and worthwhile life is one that is well above average. A bad one is one well below.

The Christian evaluation of earthly lives is misguided because it adopts a quite unjustifiably high standard. Christianity singles out the major shortcomings of our earthly existence: there is not enough happiness; there is too much suffering; the good and bad points are quite unequally and unfairly distributed; the underprivileged and underendowed do not get adequate compensation; it lasts only a short time. It then quite accurately depicts the perfect or ideal life as that which does not have any of these shortcomings. Its next step is to promise the believer that he will be able to enjoy this perfect life later on, and then it adopts as its standard of judgment the perfect life, dismissing as inadequate anything that falls short of it. Having dismissed earthly life as miserable, it further damns it by characterizing most of the pleasures of which earthly existence allows as bestial, gross, vile, and sinful, or alternatively as not really pleasurable.

This procedure is as illegitimate as if I were to refuse to call anything tall unless it was infinitely tall, or anything beautiful unless it was perfectly flawless, or anyone strong unless he was omnipotent. Even if it were true that there was available to us an afterlife that was flawless and perfect, it would still not be legitimate to judge earthly lives by this standard. We do not fail every candidate who is not an Einstein, and if we do not believe in an afterlife, we must, of course, use ordinary earthly standards.

I have so far only spoken of the worthwhileness, only of what a person can get out of a life. There are other kinds of appraisals. Clearly, we evaluate people's lives not merely from the point of view of what they yield to the persons that lead them, but also from that of other people on whom these lives have impinged. We judge a life more significant if the person has contributed to the happiness of others, whether directly by what he did for others, or by the plans, discoveries, inventions, and work he performed. Many lives that hold little in the way of pleasure or happiness for its owner are highly significant and valuable, and deserve admiration and respect on account of the contributions made.

It is now quite clear that death is simply irrelevant. If life can be worthwhile at all, then it can be so even though it is short, and if it is not worthwhile at all, then an eternity of it is simply a nightmare. It may be sad that we have to leave this beautiful world, but it is so only if and because it is beautiful, and it is no less beautiful for coming to an end. I rather suspect that an eternity of it might make us less appreciative, and in the end it would be tedious.

It will perhaps be objected now that I have not really demonstrated that life has a meaning, but merely that it can be worthwhile or have value. It must be admitted that there is a perfectly natural interpretation of the question, "What is the meaning of life?" on which my view actually proves that life has no meaning. I mean the interpretation discussed in the second section of this chapter, where I attempted to show that, if we accept the explanations of natural science, we cannot believe that living organisms have appeared on earth in accordance with the deliberate plan of some intelligent being. Hence, in this view, life cannot be said to have a purpose, in the sense in which manufactured things have a purpose. Hence, it cannot be said to have a meaning or significance in that sense.

However, this conclusion is innocuous. People are disconcerted by the thought that *life as such* has no meaning in that sense only because they very naturally think it entails that no individual life can have meaning either. They naturally assume that *this* life or *that* can have meaning only if *life as such* has meaning. However, it should by now be clear that your life and mine may or may not have meaning (in one sense) even if life as such has none (in the other). Of course, it follows from this that your life may have meaning whereas mine has not. The Christian view guarantees a meaning (in one sense) to every life the scientific view does not (in any sense). By relating the question of the meaningfulness of life to the particular circumstances of an individual's existence, the scientific view leaves it an open question whether an individual's life has meaning or not. It is, however, clear that the latter is the important sense of "having a meaning." Christians, too, must feel that their lives are wasted and meaningless if they have not achieved salvation. To know that even such lost lives have a meaning in another sense is no consolation to them. What matters is not that life should have a guaranteed meaning, whatever happens here or hereafter, but that, by luck (Grace) or the right temperament and attitude (Faith) or a judicious life (Works) a person should make the most of his life.

"But here lies the rub," it will be said. "Surely, it makes all the difference whether there is an afterlife. This is where morality comes in." It would be a mistake to believe that. Morality is not the meting out of punishment and reward. To be moral is to refrain from doing to others what, if they followed reason, they would not do to themselves, and to do for others what, if they followed reason, they would want to have done. It is, roughly speaking, to recognize that others, too, have a right to a worthwhile life. Being moral does not make one's own life worthwhile; it helps others to make theirs so.

Conclusion

I have tried to establish three points:

1. That scientific explanations render their *explicanda* as intelligible as prescientific explanations; they differ from the latter only in that, having testable implications and being more precisely formulated, their truth or falsity can be determined with a high degree of probability;
2. That science does not rob human life of purpose, in the only sense that matters, but, on the contrary, renders many more of our purposes capable of realization;
3. That common sense, the Christian world view, and the scientific approach agree on the criteria but differ on the standard to be employed in the evaluation of human lives; judging human lives by the standards of perfection, as Christians do, is unjustified; if we abandon this excessively high standard and replace it by an everyday one, we have no longer any reason for dismissing earthly existence as not worthwhile.

On the basis of these three points, I have attempted to explain why so many people come to the conclusion that human existence is meaningless and to show that this conclusion is false. In my opinion, this pessimism rests on a combination of two beliefs, both partly true and partly false: the belief that the meaningfulness of life depends on the satisfaction of at least three conditions, and the belief that this universe satisfies none of them. The conditions are, first, that the universe is intelligible, second, that life has a purpose, and third, that all people's hopes and desires can ultimately be satisfied. It seemed to medieval Christians and it seems to many Christians today that Christianity offers a picture of the world that can meet these conditions. To many Christians and nonChristians alike it seems that the scientific world picture is incompatible with that of Christianity, therefore, with the view that these three conditions are met, and therefore, with the view that life has a meaning. Hence, they feel that they are confronted by the dilemma of accepting either a world picture incompatible with the discoveries of science or the view that life is meaningless.

I have attempted to show that the dilemma is unreal because life can be meaningful even if not all of these conditions are met. My main conclusion, therefore, is that acceptance of the scientific world picture provides no reason for saying that life is meaningless, but on the contrary every reason for saying that there are many lives that are meaningful and significant. My subsidiary conclusion is that one of the reasons frequently offered for retaining the Christian world picture, namely, that its acceptance gives us a guarantee of a meaning for human existence, is unsound. We can see that our lives can have a meaning even if we abandon it and adopt the scientific world picture instead. I have, moreover, mentioned several reasons for rejecting the Christian world picture:

1. The biblical explanations of the details of our universe are often simply false;
2. The so-called explanations of the whole universe are incomprehensible or absurd;
3. Christianity's low evaluation of earthly existence (which is the main cause of the belief in the meaninglessness of life) rests on the use of an unjustifiably high standard of judgment.

Notes and References

[1]L. Tolstoy (1940) "A Confession," reprinted in *A Confession, The Gospel in Brief, and What I Believe,* The World's Classics, No. 229. Geoffrey Cumberlege, London, UK.

[2]*See,* e.g., E. R. Bevan (1932) *Christianity,* Butterworth, London, UK, pp. 211–227. *See also* H. J. Paton (1955) *The Modern Predicament.* George Allen and Unwin Ltd., London, UK, pp. 103–116, 374.

[3]*See,* for instance, L. E. Elliott-Binns (1952) *The Development of English Theology in the Later Nineteenth Century.* Longman, Green & Co., London, UK, pp. 30–33.

[4]*See* Plato, Phaedo, in *Five Dialogues,* Everyman's Library No. 456, E. P. Duton, New York, NY, p. 189, para. 99.

[5]Leibniz (1934) On the ultimate origination of things, in *The Philosophical Writings of Leibniz* (M. Morris, trans.), Everyman's Library No. 905, E. P. Duton, New York, NY, p. 32–41.

[6]*See* Leibniz (1934) Monadology, in *The Philosophical Writings of Leibniz* (M. Morris, trans.), Everyman's Library No. 905, E. P. Duton, New York, NY, pp. 8–10, para. 32–38.

[7]To borrow the useful term coined by Professor D. A. T. Gasking of Melbourne University.

[8]*See,* e.g., J. J. C. Smart (1957) The existence of God, reprinted in *New Essays in Philosophical Theology* (A. Flew and A. MacIntyre, eds.) S.C.M. Press, London, UK, pp. 35–39.

[9]That creation out of nothing is not a clarificatory notion becomes obvious when we learn that "in the philosophical sense" it does not imply creation at a particular time. The universe could be regarded as a creation out of nothing even if it had no beginning. *See,* e.g., E. Gilson (1957) *The Christian Philosophy of St. Thomas Aquinas.* Victor Gollancz Ltd., London, UK, pp. 147–155; and E. L. Mascall (1956) *Via Media.* Longmans, Green & Co., London, UK, pp. 28 ff.

[10]In what follows, I have drawn heavily on the work of Ryle and Toulmin. *See,* for instance, G. Ryle (1949) *The Concept of Mind.* Hutchinson's University Library, London, UK, pp. 56–60, and so on, and his article, If, so, and because, in *Philosophical Analysis,* Cornell University Press, Ithaca, NY (1950) (Max Black, ed.); and S. E. Toulmin (1953) *Introduction to the Philosophy of Science.* Hutchinson's University Library, London, UK.

[11]*See* references listed above.

[12]L. Wittgenstein (1922) *Tractatus Logico-Philosophicus.* Routledge & Kegan Paul Ltd., London, UK, Sect. 6.44–6.45.

[13]L. Wittgenstein, *Tractatus Logico-Philosophicus,* p. 46. *See also* R. Otto (1952) *The Idea of the Holy.* Geoffrey Cumberlege, London, UK, esp. pp. 9–29.

[14]Contemporary theologians would admit that it cannot be proved that the universe must have had a beginning. They would admit that we know it only through revelation. (*See* Note No. 9) I take it more or less for granted that Kant's attempted proof of the Thesis in his "First Antinomy of Reason" (I. Kant [1950] *Critique of Pure Reason* [N. K. Smith, trans.], Macmillan and Co. Ltd., London, UK, pp. 396-402) is invalid. It rests on a premise that is false: that the completion of the infinite series of succession of states, which must have preceded the present state if the world has had no beginning, is logically impossible. We can persuade

ourselves to think that this infinite series is logically impossible if we insist that it is a series that must, literally, be completed, for the verb "to complete," as normally used, implies an activity that, in turn, implies an agent who must have begun the activity at some time. If an infinite series is a whole that must be completed, then, indeed, the world must have had a beginning, but that is precisely the question at issue. If we say, as Kant does at first, "that an eternity has elapsed," we do not feel the same impossibility. It is only when we take seriously the words "synthesis" and "completion," both of which suggest or imply "work" or "activity" and, therefore, "beginning," that it seems necessary that an infinity of successive states cannot have elapsed. *See also* R. Crawshay-Williams (1957) *Methods and Criteria of Reasoning.* Routledge & Kegan Paul, London, UK, App. iv.

[15]*See,* e.g., Is life worth living? (1950; BBC Talk by the Rev. John Sutherland Bonnell), in *Asking Them Questions,* Third Series (R. S. Wight, ed.), Geoffrey Cumberlege, London, UK.

[16]*See,* e.g., R. Otto (1952) *The Idea of the Holy,* Geoffrey Cumberlege, London, UK, pp. 9–11. *See also* C. A. Campbell (1957) *On Selfhood and Godhood.* George Allen & Unwin Ltd., London, UK, p. 246; and H. J. Paton (1955) *The Modern Predicament,* McMillan, New York, NY, pp. 69–71.

[17]For a discussion of this issue, *see* the eighteenth-century controversy between Deists and Theists, for instance, in L. Stephen (1902) *History of English Thought in the Eighteenth Century.* Smith, Elder & Co., London, UK, pp. 112–119,134–163. *See also* the attacks by J. Toland and M. Tindal on "the mysterious" in *Christianity Not Mysterious* and *Christianity as Old as the Creation, or the Gospel a Republication of the Religion of Nature,* resp., parts of which are reprinted in H. Bettenson (1967) *Documents of the Christian Church,* Oxford University Press, New York, NY, pp. 426–431. For modern views maintaining that mysteriousness is an essential element in religion, *see* R. Otto, *The Idea of the Holy,* esp. pp. 25–40, and most recently M. B. Foster (1957) *Mystery and Philosophy.* S.C.M. Press, London, UK, esp. Chs. IV. and VI. For the view that statements about God must be nonsensical or absurd, *see,* e.g., H. J. Paton, *The Modern Predicament,* pp. 119–120, 367–369. *See also* Theology and falsification (1955), in *New Essays in Philosophical Theology* (A. Flew and A. MacIntyre, eds.), S.C.M. Press, London, UK, pp. 96–131.

[18]Stephen Neill (1955) *Christian Faith To-day.* Penguin Books, London, UK, pp. 240–241.

[19]It is difficult to feel the magnitude of this first sin unless one takes seriously the words "Behold, the man has eaten of the fruit of the tree of knowledge of good and evil, and is become as one of us; and now, may he not put forth his hand, and take also of the tree of life, and eat, and live for ever?" Genesis iii, 22.

[20]How impossible it is to make sense of this story has been demonstrated beyond any doubt by Tolstoy in his famous "Conclusion of a Criticism of Dogmatic Theology," reprinted in *A Confession, The Gospel in Brief, and What I Believe,* no. 229, The World's Classics (1940), Geoffrey Cumberlege, London, UK.

[21]*See* The Nicene Creed, The Tridentine Profession of Faith, The Syllabus of Errors, reprinted in *Documents of the Christian Church,* pp. 34, 373, and 380, resp.

[22]*See,* e.g., J. S. Whale, *The Protestant Tradition,* Ch. IV., esp. pp. 48–56.

[23]*See* J. S. Whale, *The Protestant Tradition,* Ch. IV., p. 61 ff.

[24]*See* "The Confession of Augsburg," esp. Articles II., IV., XVIII., XIX., XX.; "Christianae Religionis Institutio," "The Westminster Confession of Faith," esp. Articles III., VI., IX., X., XI., XVI., XVII.; "The Baptist Confession of Faith," esp. Articles III., XXI., XXIII., reprinted in *Documents of the Christian Church,* pp. 294 ff., 298 ff., 344 ff., 349 ff.

[25]*See,* e.g., his *An Empiricist's View of the Nature of Religious Belief* (Eddington Memorial Lecture).

[26]*See,* e.g., the two series of Gifford Lectures most recently published: H. J. Paton, *The Modern Predicament,* pp. 69 ff.; and C. A. Campbell, *On Selfhood and Godhood,* pp. 231–250.

[27]R. Otto, *The Idea of the Holy,* p. 9.

[28]F. Dostoyevsky (1953) *The Devils.* The Penguin Classics, London, UK, pp. 613,614.

[29]L. Tolstoy, *A Confession, The Gospel in Brief, and What I Believe,* p. 24.

[30]*See,* for instance, J. S. Whale, *Christian Doctrine,* pp. 171, 176–178, and so on. *See also* S. Neill, *Christian Faith To-day,* p. 741.

The Meaning of Life in Old Age

Harry R. Moody

At first we want life to be romantic; later,
to be bearable; finally, to be understandable.
—Louise Bogan

I approach the question of meaning in old age as a philosopher, yet not exclusively from a philosophical point of view. Alasdair MacIntyre has suggested that every philosophy presupposes a sociology, so it is just as well to be explicit about how social structure is related to ideas. If MacIntyre is right, then the examination of a seemingly remote or metaphysical question—"What is the meaning of life?"—may have extraordinary implications for how we think about the social system, about ethics and politics, and even about the daily activities of our lives. It may prove to be the key to how we can think about the problem of meaning in the last stage of life.

I begin my inquiry by trying to make clear how we can succeed in thinking about the meaning of life and I conclude that we inevitably invoke some image of life as a whole, of the unity of a human life. Contemporary psychological systems appeal to some such idea, but it is rarely made explicit. We live in a culture dominated by the therapeutic outlook, a world that looks to psychology rather than to traditional disciplines of religion or philosophy to find meaning in life. In practice, the perspective of psychological humans tends to reinforce a separation between the public and private worlds, a separation that is a dominant feature of our society.

Aging and Ethics Ed.: N. Jecker

As we trace the origins of these psychological ideas, their ancestry reaches back eventually to Greek and Roman thought. We know that in time a suppressed dimension of ancient philosophy—the appeal to a principle of divine transcendence—eventually triumphed in the form of religion. Yet both ancient and medieval civilizations took for granted that the contemplative mode of life represented the highest possibility for human existence. By contrast, the modern world, since the seventeenth century, has favored a life of activity over a life of contemplation. This fact is fundamental to understanding the modern horror of old age, which is a horror of the vacuum—the "limbo state" of inactivity.

In twentieth-century philosophy, the older problem of the meaning of life for a time disappeared, but when it resurfaced, it was assumed that the solution must lie in some form of privatism: the meaning of *my* life. Modern philosophy has rejected any appeal to transcendent sources of meaning. Further, the modernist wish is to maintain that life in old age *can* have meaning even if life as a whole does not. The traditional answer was quite different. The traditional answer amounted to what Philibert called a "scale of ages" or a hierarchy of life stages in which late life was a time for unfolding wisdom and spiritual understanding. However, any such positive image of old age depends on a cultural framework wider than the individual.

That cultural framework is what is missing today. The modernization of consciousness coincides with the modernized life cycle, with its sharp separation among the "three boxes of life" (education, work, retirement). This segmented life course undercuts any sense or meaning that belongs to the whole of life. The search for meaning is displaced from otherworldly to this-worldly concerns, then finally compressed into late life, and brought under the domination of professions and bureaucracy. The result has been a covert ideology of lifespan development that lacks any rational foundation for shared public values whereby the idea of development might make sense. In the end, the empirical science of lifespan development must turn to the humanities and to cultural traditions if we seek to reconstruct a narrative unity to the human life course.

As soon as we begin to explore the question of meaning in old age, we come up against an obstacle presented by the forms of thought that prevail in our time. How are we to understand "meaning" in life? It is characteristic of modern thought to separate three levels of meaning: the individual, the collective, and the cosmic. That separation is what defines the present situation and throws up an obstacle to the inquiry.

That separation has a history of its own. After the Enlightenment, the cosmic sense of meaning began to atrophy. In its place came belief in a collective sense of meaning that absorbed into itself both the individual and the cosmic senses of meaning: the idea of progress through history. Then in the twentieth century, the collective sense of meaning, in turn, has weakened, leaving us with an exclusively individual preoccupation with the meaning of life. Religious and metaphysical systems have lost credibility, belief in social justice and human progress has become doubtful, and the best advice appears to be Voltaire's prescription: "Cultivate your garden." In other words, retreating to a private sphere of meaning is the best we can hope for in a disordered and meaningless world. We have gone the way of Candide: we can hope at best to cultivate our gardens, but those gardens have now shrunk to the size of windowboxes.

This retreat to privatism is an unsatisfactory solution to questions about the meaning of life and the meaning of old age. Privatism, what Christopher Lasch has called "the culture of Narcissism," is unsuited to provide any abiding sense of meaning that transcends the individual life course. That absence of enduring meaning creates a special peril for old age when the temptations of narcissistic absorption are greater. The psychology of narcissism—the self-reflexive center of psychological motivation—has emerged as a dominant problem for modern individuals.[1] The heart of the problem is that, in the modern age, at just the moment when the meaning of old age is in question, the cultural fabric itself unravels to the point where "the center cannot hold." Things fall apart and the search for meaning turns inward to escape from collapsing institutional norms, but in a therapeutic culture, can late-life narcissism find satisfactory resolution apart from any binding institutional norms?[2]

The bankruptcy of privatism becomes all too evident in old age, but by that point, individuals alone cannot invent meanings to save themselves from despair. The exercise itself invites the humor of Samuel Beckett and others who have seen the problem for what it is. We have shrunk the question of meaning down to its lowest denominator, the psychological meaning of *my* life, but it turns out that any serious inquiry, even in that direction, brings us quickly to a wider collective and historical level: the survival and meaning of the human species.

Finally, we cannot consistently sustain a sense of meaning without situating our collective enterprise in a wider cosmic setting. The philosophical questions are all connected, regardless of how we come to answer them. The levels of meaning—individual, collective, and cosmic—must be connected. As we become increasingly an aging society, the collapse of a coherent framework for meaning in old age becomes a more pressing social and cultural problem. It cannot be resolved without clarification of the philosophical issues, and it is to these issues that I now turn.

I have suggested that we can distinguish between three levels of meaning in life: the meaning of my life, the meaning of human life, and the meaning of the cosmos. In each case, we could add the phrase "as a whole." Ordinarily, human beings act as if they understood quite well the range of meanings that life can in fact possess, but when we add the qualifying phrase, "life as a whole," then we are clearly talking about something else, something out of the ordinary. It is this larger or global sense of meaning that comes to attention at times of crisis and, particularly, when the *limits* of a life come into view: for example, in the shock that surrounds death and bereavement. These moments, which Jaspers called "limit situations" *(Grenzsituationen),* are the familiar signposts of modern literature and existential philosophy. Characteristically, in literary expression or philosophical reflection, as soon as we begin to reflect on the meaning of life *as a whole,* we quickly reach the boundaries of language itself, as Wittgenstein noted: "The solution of the problem of life is seen in the vanishing of the problem. Is this not the reason why those who have found after a long period of doubt that the meaning of life *(Sinn des Lebens)* became clear to them have then been unable to say what constituted that meaning?"[3] However, contrary to Wittgenstein's contention, the problem of the meaning of life has not vanished. On the contrary, with the emergence of an aging society, it seems certain on demographic grounds alone that more and more people will be confronted with the "limit situation" of old age in which the whole of life itself may be put into question. My purpose here is to examine how that question is intelligible, in both cultural and philosophical terms.

At certain times in life, we tend to ask questions about the meaning of life *as a whole.* In midlife crisis or in autobiographical reflection in old age, what is at stake is a sense of the meaning (or lack of meaning) of my life.[4] This is the psychological version of the question of meaning that is most familiar today, but we must also distinguish a second sense of this problematic question about the meaning of life. This is the concern about the meaning of the entire human situation, the meaning of human existence or human history *as a whole.* This concern is alive today in fears about the threat of nuclear war and the future of the human species. Here, obviously, is something more than individual concern alone. For old age, that collective concern can be felt as disillusionment with the collective goals and efforts of earlier years. All earlier goals presupposed an image of the future,[5] but in old age, that imagined future has now become history. The results are likely to be different from what was anticipated, and effort often gives way to disillusionment. Struggles to create a more just world, for example, run up against a mood expressed in Ecclesiastes: "I saw that under the sun the race is not to the swift, nor the battle to the strong, nor bread to the wise, nor riches to the intelligent, nor favor to the men of skill; but time and chance happen to them all." The pessimism of Ecclesiastes is not uncommon in old age. In our time, doubts about progress and the advance of social justice have become endemic. In old age, this questioning about the meaning of human life as a whole may be unavoidable.

Finally, we come to the widest possible question about meaning in life: the meaning of human existence in the cosmos, the meaning of life and death. These are the questions of traditional philosophy and religion. The questions may be greeted by the cry of Ecclesiastes, "All is vanity;" or by faith in an inscrutable reality beyond understanding, a faith expressed for example in the Book of Job; or again by belief, as we find in works by Aristotle and Spinoza, that human existence is part of an order of nature that makes our human life intelligible at a cosmic scale. Note that I have not stressed here the question of belief in immortality or life after death. This is, of course, a vital question, but I am concerned at the moment to bring out alternative responses to what is a more fundamental question about *cosmic* meaning: Does our human existence have an ultimate significance in the universe as a whole?[6] This is a level of meaning wider than either individual or collective concern. Nonetheless, we can presume that any answer to this wider question of the meaning of the cosmos or of human life as a whole will have implications for the psychological or personal question of the meaning of *my* life.

Any question about the meaning of life finally comes back to intuitions about wholeness: the unity or wholeness of an individual life, the unity of the human species, the unity of the universe as a whole. As we approach the completion of any task—building up a new business, writing a book, leaving one job for another—it is inevitable that we think of the task *as a whole.* We ask the question, "What did it mean?" It is as if we were on a ship or a plane leaving a city and then, at a certain distance, we looked back and saw the city as a whole for the first time.

However, the analogy fails, as Kant noted in his *Critique of Pure Reason.* When I ask questions about the unity or totality of myself or my life, I am still within my life, not outside it, not able to see it as a whole. The same point holds true for the human species or human history, in which we live our lives, and again, at the widest possible scale, for the universe as a whole. Thus, argues Kant, it is a logical error to think that we can form intelligible concepts of the self or of the universe as a whole, but we cannot resist the impulse to seek such concepts, despite the fact that we can never grasp the metaphysical totality of things.[7]

Kant's critical observation opens up a different kind of question about the search for meaning over the course of human life. The question is transformed into a phenomenological inquiry into the different forms of meaning. How are the three levels of meaning—individual, collective, and cosmic—bound up with one another? What is the shape of this question about the meaning of life in old age, and how is the question related to the status and social meaning of old age itself? How have these relationships changed historically over time, and what are the possibilities for rediscoving old age as a period of life in which the search for meaning has a legitimacy and even a broader social importance?

The question of meaning in old age, after all, is not merely an academic inquiry; it has implications for the quality of life in old age in our world today. Old age is the period of life when the shape of life as a whole comes into view, or when it is natural, at any rate, to try to see things in a wider scale. Totality may be metaphysically unattainable but the drive toward totality, the search for meaning, appears at a point when the task of life is about to be completed. In Hegel's phrase, "The owl of Minerva takes flight as the shades of dusk are falling."

Is it permissible to use the word "task" in speaking of this search for meaning? Even lifespan development psychology cannot avoid the language of developmental "tasks" for different life stages. It is just here that old age is in a precarious position, for the task of the final stage of life is in some way bound up with the *completion* of life as a whole. Old age is not simply one more stage but the final stage, the stage that sums up all that went before. Here, we cannot avoid thinking of life *as a whole,* but can we speak of a task for life as a whole without some larger philosophical concept of meaning? Also, without this larger intuition about life as a whole, what sustains the integrity of the last stage of life?

Inevitably, we find ourselves caught up in an image of the *normative* life cycle, but what the social sciences have lacked is a philosophical legitimation for such covert appeals to the idea of a normative life cycle. In the work of our most prominent exponent of such a theory—Erik Erikson—such covert ethical appeals command widespread public admiration.[8] Erikson's ideals evoke in us a numinous image of something we desperately want to believe in: the unity of the life course, the integrity of the life cycle. In Erikson's work, developmental psychology becomes a vehicle of hope.

Yet at the crucial moment, the moment when we need to justify the purposes that give meaning to life, the psychology of lifespan development falls apart conceptually. Fundamental questions are left unanswered. How are life stages, after all, to be distinguished from one another? What about the contrasting roles of fate and freedom in the developmental cycle? By what sleight of hand do we deduce values from the empirical data of psychological change?

The implications of this conceptual incoherence are, unfortunately, quite serious. The failure to articulate a philosophical notion of meaning across the lifespan means that practical activities, such as psychotherapy, which are based on normative theories of lifespan development, cannot be fully successful. I believe, too, that questions of meaning are of the utmost importance for practical decisions in ethics and public policy. Without reflection on these metaphysical questions we will inevitably lose our direction when we try to think about very specific matters in gerontology: for example, the ethics of suicide and euthanasia, cognitive development over the lifespan, reminiscence and life review, late-life education, adult counseling and psychotherapy. Practice and theory both depend on certain shared normative grounds. The attempt to cut loose the problem of meaning from its philosophical presuppositions cannot succeed. Our concepts of lifespan development must instead retrace the path through the philosophical tradition out of which they emerged.

When we speak of *meaning* of life in old age, we may intend several very different things. These multiple meanings become clear as we retrace the history of the search for meaning in the philosophical tradition. Questions about the meaning of life reach back to the earliest period of Greek philosophy. It was Socrates who framed the question in concise form: "The unexamined life is not worth living." Socratic questioning in turn found an answer in the Platonic doctrine of man and the cosmos. For Plato, the answer was *wholeness:* the balanced functioning of all the powers of the human soul, powers guided by the light of reason reflecting a divine cosmos. The Platonic answer was matched by a system of education for lifespan development where each stage of life would prepare the way for the final vision.

For Aristotle, too, integration of self over time was the touchstone. Aristotle's term *eudaimonia* is variously translated as "happiness," "self-realization," or, in a more global context, "living well."[9] The Aristotelian view sees a good life in old age as a culmination of dynamic elements operating over an entire lifespan: One is happy when one has the good fortune to have access to all those elements (friends, sufficient wealth, good character) required for living well. The *meaning* of old age, then, would be found in just that immanent pattern of activity that allows the human being to exercise those powers that belong to human nature.

Viewed in this naturalistic fashion, the metaphysical questions of the meaning of life cannot be separated from the natural structure of the entire lifespan. It is not isolated moments of pleasure, but the fulfillment of an entire course of life that determines whatever happiness or meaning a life can have. Consequently, happiness in old age remains precarious. "Call no man happy until he is dead," the Greeks said. In other words, meaning is contingent on the circumstances of life. Beyond this, we are compelled to say, on these premises, the question of the meaning of life cannot even arise. To

Aristotle the question would not even make sense. The Aristotelian understanding of meaning in old age is, thus, not far from what Freud meant when he remarked that the moment a person begins to question the meaning of life, he is already sick.

The difficulty with the classical Aristotelian account of virtue and *eudaimonia* is that it appears to make human fulfillment excessively dependent on the force of external circumstance. That limitation was the point of departure for the philosophical movement known as Stoicism, which took its inspiration not from Aristotle, but from the ethical example of Socrates. For the Stoics, the meaning of a human life could be understood only as self-integrity and self-possession: living in accordance with one's nature. How can this ethical self-integrity be reconciled with the loss of self entailed by death?[10]

It is a fact of our nature that we grow old and die. From the Stoic point of view, then, the natural human life cycle must be accepted and lived out in full self-possession. Each stage of life has its own integrity and, hence, its own meaning. Such a stance enables us to bear the existential pain of aging: the facts of perishing, loss, the disappearance of the past. How are these losses to be reconciled with the goal of living well in late life? It is here that the Stoic view of the role of pain and suffering takes on its importance. Rational suicide remains one possibility, but the supreme ethical demand is always for rationalistic self-control.[11]

The Stoics were the first philosophers to offer a coherent response to the philosophical problem of aging. The most influential account along these lines appeared in Cicero's *De Senectute,* where old age acquires a meaning identified with the achievement of total self-possession, ego-integrity, and wisdom. In our psychologically oriented culture, the terms "ego-integrity" and "wisdom" evoke immediately association with Erik Erikson's Eight Ages of Man.[12] I have given this picture of the Stoic perspective on aging precisely to stress the ways in which Erikson's own psychology, on its normative site, is finally only a restatement of Stoic ideals. In the last stage of life, in old age, Erikson finds a distinctive polarity between ego-integrity and despair. To resolve that polarity, and to achieve wisdom, argues Erikson, constitutes the task of the last stage of life. Erikson's virtue of ego-integrity, like Freud's courage in the face of suffering during his last illness, reveals that behind psychoanalytic theory stands an ancestry of Stoic philosophy.

The psychoanalytic approach to the meaning of old age is a naturalistic view of human fulfillment. It is a balanced view in that there is no room for overconfidence, nor is there any recourse to pessimism or appeal to divine grace. Reason, or ego-strength, belongs to our nature just as much as instinctual conflict. To find a balance between reason and instinct promises a state of equilibrium or "living in accordance with our nature," which is the opposite of repression. Achieving this wisdom of self-possession may require a lifetime of discipline, but wisdom alone retains the strength to offset the inevitable losses of old age without retreat to narcissism or despair. Old age, then, in both the Stoic and psychoanalytic view, is not the metaphysical "completion of being" but simply the final test of the human being's hold on the reality principle.

In the Roman world, the lofty idealism of Stoic ethics was felt by some to be unsatisfying, in part because it seemed to enclose human consciousness too narrowly within the boundaries of all-controlling reason, but the Stoic ideology of self-contained consciousness also embodied a deeper despair about the integrity of the wider social order. The Roman concept of citizenship had decayed. For the elite who were attracted to Stoic ideas, the sphere of meaning had shrunken to a smaller, more manageable scale. Transcendence became impossible; self-mastery was everything, but the claims of transcendence would not remain long suppressed. We know that the historical outcome of the conflict of ideas in antiquity was the triumph of a religious view with the coming of Christianity. For a thousand years and more, the symbols of transcendence would provide an answer to the meaning of life.

Transcendence can mean many different things. Peck, speaking of old age, has pointed to a polarity between ego-preoccupation and ego-transcendence. Ego-preoccupation represents the temptations of narcissism in old age: withdrawal of feeling for others or for concerns that go beyond individual or personal existence. By contrast, ego-transcendence means living for the sake of causes or objects that lie beyond the self.[13]

Peck speaks of ego-transcendence in a purely descriptive, psychological sense, but the object or cause of transcendence can be understood in either humanistic or religious terms. In both cases, the object is a source of meaning that allows me to make sense of my life because that object is what survives my death. Ego-transcendence may be understood either as a kind of secular immortality or as faith in a transpersonal cosmic reality (salvation).

The concept of transcendence as secular immortality is discussed by Lifton,[14] where it signifies the capacity of the individual ego to project its concerns onto a social institution or aesthetic object that survives individual extinction and thereby confers meaning on the temporal striving of finite individual existence. This level of meaning corresponds to the collective, historical sense of meaning described earlier. My life means something because I am part of a human enterprise larger than myself.

In addition to this humanistic sense of transcendence, one can recognize the search for meaning as an encounter with what transcends not only the finite ego, but even the human history itself. It is this encounter with absolute Reality that the historical religions speak of in the language of salvation or deliverance: *moksha, samadhi, 'irfan,* or other terms denoting the experience of transcendence.[15] The language has its origin in the experience of a numinous reality encountered in contemplative consciousness.[16] Indeed, cosmic transcendence cannot be understood apart from the experience of contemplation or detachment from the world of activity.

Cultures differ in their appreciation of the virtues of contemplation versus activity, but there is no question that modern culture prizes activity to the degree that the idea of contemplative receptivity has become almost unintelligible.[17] This failure to understand contemplation poses a grave problem in the process of growing old today. Without some feeling for the virtues of silence, inwardness, patience, and contemplation, it is impossible for us to understand what ego-transcendence in old age might ultimately be about. We can only form a distorted image of it and call it "quietism" or "disengagement." The next move is then to retreat to secular immortality (i.e., living on through our actions) as a version of transcendence more acceptable to the activist temperament of modernity.

A vigorous defense of an activist style of growing old is offered by Simone de Beauvoir in *The Coming of Age,* where she warns about what must be done in order to escape the existential vacuum of late life:

> The greatest good fortune, even greater than health, for the old person is to have his world still inhabited by projects: then, busy and useful, he escapes both from boredom and from decay. ...There is only one solution if old age is not to be an absurd parody of our former life, and that is to go on pursuing ends that give our existence a meaning—devotion to individuals, to groups or to causes, social, political, intellectual or creative work. In spite of the moralists' opinion to the contrary, in old age we should wish still to have passions strong enough to prevent us turning in upon ourselves.[18]

Old age, in this view, is not a time for wisdom or summing up. It is a time for continual engagement.

Simone de Beauvoir's existential view of old age, although perhaps extreme, is not really far from the view that prevails among the enlightened upper middle classes who articulate the dominant values in our society.[19] The style of activist aging seems to embody our preferred solution to the problem of aging. It is a widely shared image of ideal old age and is entirely coherent with dominant social institutions and with the activist ethos of the modern age itself.[20] Here, a remark by Schuon is appropriate:

> According to an Arab proverb which reflects the Moslem's attitude to life, slowness comes from God and haste from Satan, and this leads to the following reflection: as machines

> devour time modern man is always in a hurry, and, as this perpetual lack of time creates in him reflexes of haste and superficiality, modern man mistakes these reflexes—which compensate corresponding forms of disequilibrium—for marks of superiority...[21]

It is common for young or middle-aged people today to feel exactly this sense of superiority toward the elderly: old people are slow-moving, not modern, even throwbacks to an earlier era. So we can easily feel a sentimental pity toward the old when they are unable or unwilling to share those reflexes of haste and superficiality that have become our daily habits. In our common desire to help the elderly, what we secretly wish is to prolong the haste that excludes us from even a moment of quietness and contemplation. We find the quietness of the old, even their very presence, disturbing, as if it were a repudiation of all that we hold dear. Old age, like death, is an indictment of that fantasy of agitation that the young and middle-aged take to be the meaning of life. Old age appears only as a limbo state, an absence of meaning in life.

Activity as a refuge from the limbo state of aging: this formula is the preferred solution to the problem of late-life meaning in the modern world. A cursory glance at the academic literature of gerontology will turn up the familiar advice that we find in the popular literature of the organized old age associations, such as the American Association of Retired Persons' magazine *Modern Maturity*. The formula is always the same: keep busy, keep active. This formula is both a sociological statement about institutional life and a philosophic world view. What we now need is a philosophical critique of how this formula came to achieve its almost hypnotic power over the culture at large. To carry out that critique requires some further examination of philosophy and meaning. It will be clear that philosophy in our time has, in fact, defined a point of view about the meaning of life that is completely coherent with the dominant institutions of our society. This philosophical point of view must now be the center of our inquiry.

Twentieth-century philosophy has offered an ambivalent response to questions about the meaning of life. On the one hand, existential philosophers such as Heidegger and Jaspers have pointed to features of modern society that prompt individuals to experience their lives as meaningless.[22] In this view, old age is the stage in life when individuals are forced to confront an existence that has been meaningless all along. Old age is the time when we discover the emptiness and self-deception behind goals and values that were taken for granted. The central preoccupation of existential philosophy, broadly understood, is to understand how it is possible for human beings to live an authentic existence in a world where traditional meanings are no longer convincing. Although existential thinkers differ in their answers, they agree on the importance of the question of the meaning of life.

By contrast, Anglo-American analytic philosophers for many years dismissed the question of the meaning of life as illegitimate. An extreme point of view, developing largely under the influence of logical positivism, held that the phrase "the meaning of life" was meaningless. Following a line of critical thought initiated by Kant, questions reaching beyond empirical knowledge were viewed as attempts at metaphysics, but in recent years, this hostility to metaphysical questions has diminished and become less doctrinaire. At the same time, there are few who are willing to return to traditional underpinnings of confidence in the meaning of life. Thus, the philosophical tendency is to reaffirm traditional questions of global meaning while avoiding the traditional answers, generally religious or metaphysical, that have been given to the question of the meaning of life.

Recently, this question has again emerged as important.[23] In the first half of the twentieth century, analytic philosophers banished the question of the meaning of life from rational discourse. In recent years, some prominent analytic philosophers[24] have suggested that questions about the meaning of life are intelligible and, in fact, important. Finally, a few philosophers[25] have questioned whether academic philosophical thinking is itself somehow debarred from coming to grips with the full existential depth of the question of ultimate meaning in life.

The force of philosophical argument in our time is to drive a wedge between meaning for an individual life and any wider sense of meaning that could have a rational foundation. The connection between the two levels appears increasingly arbitrary, thus reinforcing the split between the public and private worlds. Modern thought characteristically maintains that objective elements of meaning beyond the self—the meaning of life as a whole—are not a necessary or sufficient condition for the subjective sense of meaning, the meaning of the individual life. In other words, modern thought tends to insist that, even if the universe as a whole or human history as a whole lacks meaning, nonetheless, my life can have a meaning.

Existentialists express this idea in a mood of radical anxiety and see it as the necessity to choose, once and for all, the final meaning of our lives in every one of our actions. By contrast, ordinary people go about their lives pursuing projects and hobbies that give their lives meaning in a purely private sense. That private pursuit of meaning today is often irreligious, and increasingly often, people lack confidence in any grand historical design. Perhaps, in the age of nuclear weapons, they even lack faith in the collective future of humankind. In short, neither the universe as a whole, nor human history as a whole, inspires confidence in total or objective meaning.[26] Consequently, people go about their business pursuing a form of meaning that is increasingly privatized. In old age, as time runs out, the final question remains to be asked: what did it all mean?

Old people commonly encounter this question in moments of autobiographical reflection and reminiscence about the past. This fact has given rise to the growing literature surrounding the concept of "life review," originally introduced by Butler[27] as interpretation of the phenomenon of reminiscence in old age, but the appeal of the idea of life review does not come from its scientific or explanatory status. The vogue among gerontologists for life review ("reminiscence therapy") and oral history springs from many sources. At bottom, though, the attraction of these ideas represents a wish to find in the experience of old people elements of strength and positive affirmation. Yet this positive impulse bears in it a danger of degenerating into mere wish-fulfillment and sentimentality, where the concept of life review becomes a kind of ersatz spirituality in humanistic costume.

Butler speaks of the life review as an effort by the older person to sum up an entire life history, to sift its meaning, and ultimately to come to terms with that history at the horizon of death. The gerontological literature on life review makes it clear that a wide variety of professionals in gerontology identify the life review as an opportunity for the elderly to find meaning in life through autobiographical consciousness.[28] In sum, the normative underpinning of life review involves an appeal to ethical, philosophical, and even spiritual values. It incorporates a specific view, recast into psychological language, of where meaning in old age will be found.

Are there standards, then, or criteria, by which a life review could be judged successful? How are we to know whether an act of reminiscence is a discovery of final meaning or perhaps just another "metaphor of self" constructed for obscure purposes, or even a new form of self-deception? It is here that the philosophical literature turns out to be illuminating. A good example of the style of recent philosophical thought is the essay by Kurt Baier, "The Meaning of Life."[29] In that discussion, Baier attempts to analyze and refute the widespread conclusion that human existence is meaningless: a conclusion supported, in different ways, by both existentialist and positivist thinking. However, this pessimistic conviction of meaningless existence, Baier argues, arises from too strict a standard in the first place. That strict standard holds that human existence would be meaningful if and only if:

1. The universe is intelligible;
2. Life has a purpose; and
3. All people's hopes and desires can ultimately be satisfied.

However, Baier adopts a more modest stance toward the meaning of life: the familiar privatism or subjectivism that characterizes the modern mind and that Baier expresses so well:

> People are disconcerted by the thought that *life as such* has no meaning...only because they very naturally think it entails that no individual life can have meaning either. They naturally assume that *this* life or *that* can have meaning only if *life as such* has meaning. But it should by now be clear that your life and mine may or may not have meaning (in one sense) even if life as such has none (in the other). Of course, it follows from this that your life may have meaning while mine has not.[30]

In the act of individual life review, then, I may discover that *my* life can have a meaning even if life in general is meaningless. In fact, Baier's three conditions for meaning can be transposed from the objective or cosmic level to the level of individual autobiography. Instead of asking about the meaning of life as a whole, I may ask about the meaning of *my* life, and this latter form, in an acutely felt way, is how the question of the meaning of life is often understood in late-life reminiscence. When the question of meaning is transposed to the level of autobiography, what we have in effect is a set of conditions for inquiring into the intelligibility of life review.

Following Baier, we may suggest that a successful resolution of the old age life review signifies that:

1. My life is intelligible;
2. My life has a purpose; and
3. My hopes and desires ultimately can be satisfied.

These conditions amount to causality, teleology, and happiness. Stated in this way, we not only make transparent the philosophical presuppositions of life review, but we also link those presuppositions to older philosophical questions about the nature of causality, teleology, and happiness. The three conditions are the covert normative underpinning for life review.

Presumably, a life review that, in fact, arrived at the fulfillment of all three conditions would be successful in the sense that the person would have achieved self-knowledge or wisdom, ego-integrity, and self-actualization. What is the opposite of these? Clearly enough, lack of intelligibility, lack of purpose, lack of happiness. A person might, however, possess one or two of these three but lack a third. For example, Montaigne's later years were filled with a profound sense of self-knowledge and self-acceptance, yet the purpose of his life, as embodied in his earlier idealistic Stoicism, was largely unfulfilled.[31] In some ways, Montaigne's final philosophy is best understood as an affirmation that humans are most themselves when they live without purposes that extend too far into the future.

It is revealing that Montaigne, unlike Augustine or Rousseau for example, does not engage in life review or autobiography. He refuses temptations of totality. Instead, Montaigne writes a journal, a collection of *Essaies (essayer:* to try, to attempt, to experiment). For Montaigne, in other words, there is no summing up, no final perspective point from which to "see life steadily and see it whole." To see life at all is to see it at a particular moment, from a glance, over one's shoulder, so to speak. The wholeness of the self eludes intelligibility. Still less can the entire movement of existence be said to have a purpose in Montaigne's world view. This characteristic relativism and fluidity give Montaigne's *Essays* a quality we inescapably call modernity.

What then to we make of the search for the meaning of life at the level of individual autobiography? Simply this: that there are at least three distinct threats woven into the concept of "the meaning of my life." These are the three conditions identified earlier: that my life be understandable, that it be purposeful, and that it be happy. We can imagine a life to be happy without being either purposeful or intelligible, and vice versa. We can also imagine all three conditions obtaining at once. This last combination would be the "strong" condition for meaningfulness, but weaker conditions are also acceptable. In short, I argue that the concept of "the meaning of my life" is a multivalent concept, a set of "family resemblances" (Wittgenstein's phrase) weaving together interrelated but distinct ideas.[32] It follows that we cannot say that my life altogether lacks meaning if any single condition is unfulfilled, but what if all three conditions fail to be fulfilled? If my life lacks intelligibility, lacks a purpose, and is miserable, then it seems unavoidable to say that it is meaningless.

In alluding to Wittgenstein's concept of "family resemblances," what I suggest is that, philosophically speaking, there are a variety of "languages," "metaphors," or "world hypotheses" that constitute a plurality of conceptual paradigms for looking at the meaning of life through autobiographical consciousness.[33] Within each of these conceptual paradigms, the life review can be judged according to values of truth, authenticity, integrity, and so on, but each of the conceptual paradigms remains incommensurable with the others. Let me briefly characterize four of these paradigms: the psychological, the spiritual, the literary, and the philosophical.

The dominant presupposition of our time is the reductionism of psychology. In psychological autobiography, we reduce life review to a causal–psychological process. We see the process at work in gerontology and the human service professions of social work and psychiatry, where the concept of the old age life review was vindicated. Psychological autobiography is the taken-for-granted psychological reductionism that comes as second nature in our epoch: the "age of suspicion," as Nietzsche shrewdly called it. Psychological autobiography was dominant in the milieu where the term "life review" was first used. It remains the dominant system of interpretation. Under that species of pan-psychologism, the task of facilitating life review is easily assimilated into some variety of psychotherapy. Accordingly, the criterion for success of life review is understood to be adaptation, mental health, or some other standard of psychological functioning.

In spiritual autobiography, life review is seen as the path to salvation or deliverance. This understanding of the meaning of reminiscence goes back at least to St. Augustine's *Confessions,* which is still the paradigmatic source even for autobiographies that reject Augustinian assumptions, such as those of Rousseau or Sartre. The distinctive factor in spiritual autobiography is that an individual life story is depicted as a stage on which a universal, spiritual drama is enacted in memory. On this stage for memory, universal myth and individual historicity coincide. Like parallel lines that meet at the horizon, individual lives lose their separateness at the "still point" of eternity. Spiritual autobiography discovers in each of these parallel lives a transcendent coincidence of destiny. All differences of individual existence are finally absorbed at the horizon of divine truth. Time and memory are taken up in eternity, and the act of reminiscence becomes an act of prayer or meditation. This form of spiritual autobiography may also overlap with a psychotherapeutic understanding of life review, as in some schools of humanistic psychology where "self-actualization" substitutes as a secularized version of salvation

The literary or artistic form of autobiography arises when we see life review as a manifestation of artistic creativity. This form of the autobiographical consciousness is celebrated by James Olney as the power of metaphor,[34] but vindicating life review as a form of creative invention of the self means that the criterion of truth in autobiography loses its earlier transparency as self-discovery or self-disclosure. In modern literature, no such naive version of self-disclosure is possible. Instead, modernism is obsessed by the dialectic of sincerity and authenticity.[35] We live our lives as ongoing narratives, stories we make up as we go along through the life course. The possibility that those life stories—or "metaphors"— might have the freedom of fiction or myth-making opens up a dizzying prospect of creativity, and that prospect, in turn, leaves us looking for clues for the interpretation (hermeneutics) that must go hand in hand with the creative act of self-expression (poetics). In place of the unmediated vision of traditional self-knowledge, literary autobiography moves toward a more radical "deconstruction" of the self. We find ourselves in the landscape of postmodernism where the objectivity of self-knowledge is eclipsed in favor of pluralism. In the postmodern world, meaning is to be found only in a fictive version or metaphor of the self.[36] The duality of appearance and reality in autobiography has collapsed, to be replaced by the search for a new taxonomy of metaphors elicited from the linguistic forms in which autobiographical acts occur. Whether any transcendent form of truth can be salvaged here remains to be seen.[37]

In philosophical autobiography the status and truth-claims of the autobiographical consciousness emerge as central or problematic in their own right. Philosophical autobiography is found preeminently in those autobiographical works written as philosophical treatises or, alternatively, in first-person life stories written by great philosophers (such as Augustine, Montaigne, Rousseau, Kierkegaard, and others). These works of philosophy undercut any reductionist attempt to convert them into psychological epiphenomena, because the works incorporate in their own logic and rhetoric all the problematic philosophical questions that reductionism itself fails to address: namely, the status of time and memory, our knowledge of other minds, the intersubjectivity of language, and the nature of the self. What characterizes philosophical autobiography is that such acts of personal narrative include both questions and answers to these perennial philosophical questions.[38] A philosophical autobiography need not be autobiography written by a philosopher. Rather, it is found whenever the activity of life review turns back on itself to reflect on its own purposes and assumptions.

Psychological, spiritual, literary, and philosophical autobiographies are genres of autobiography that all have their counterparts in life review, whether written or not. The same "deep structures" of meaning are apparent in old age reminiscence as well as in refined works of autobiographical writing. The attractiveness of life review lies in its covert appeal to those same deep structures of meaning. We want the last stage of life to possess this resonance of meaning that reassures us of the integrity of the life cycle in a world where values are in flux.

Life review, the autobiographical consciousness, and the developmental psychology of life stages: these are our modern ways of structuring human time, the time between birth and death. Clearly, there are some life cycle transitions, such as infancy, maturation, puberty, senescence, and death, that are rooted in biological rhythms, but beyond this, the structure of human time is fluid and the meaning of aging is ambiguous, subject to interpretation. As long as the time of the life course is anchored in cosmic time, then the human life cycle and the cosmic cycle will, so to speak, "reverberate" with one another. Traditional *Homo religiosus* experiences time as an individual rhythm contained in a larger cosmic process: a ripple in the midst of a vast ocean of existence.[39]

For modern humans, there is no such coincidence between individual, collective, and cosmic time. The time of the human life cycle is irreparably shattered by the historical fact of modernity. That discontinuity severs the ties between these levels of meaning, and makes meaning at each of these levels problematic and open to multiple interpretations. Psychological, spiritual, literary, and philosophical autobiographies represent alternative modes of interpreting the life course, but to be committed to any one of these modes is already to take up a commitment to a certain framework of meaning. Once the framework is given, only certain questions can be asked. It is characteristic of the modern situation that these alternative commitments or interpretations of the life course lie before us, not as a given, but as matters of choice, and any choice is susceptible to infinite revision, opening up a plurality of life worlds. With that vast expansion of choices, a chasm opens up. This is the chasm between the certitude of any wider, shared meaning and the condition of open-ended individual choice that can at any time call all commitments into question. It is this chasm that constitutes the situation of modernity.[40]

Traditional societies contain dual, even contradictory images of the movement of aging: a downward movement toward debility and death and an upward movement toward unifying knowledge. The first movement is an invariant motion of organic existence, whereas the second movement represents the possibility of old age as a period of wisdom and plenitude.[41] Let me stress as strongly as possible that this traditional doctrine is not meant to be an account, even disguised, of the actual position of power or status of the aged in society, still less of any alleged veneration accorded to the aged. It is a statement about meaning, not about social structure. In Puritan America, for example, only the well-to-do elderly were venerated, and there is evidence of competition or intergenerational conflict revolving around property and power. Nevertheless, Puritan America, like other traditional societies, incorporated a keen sense of the dual movement of aging: of the dialectic between spiritual strength and physical frailty that was confirmed in the religious tradition.[42]

What is important about such traditional doctrines of old age is not how they support the social status of the aged in society, but how they furnish individuals with a cultural framework for finding meaning in their experience. In this view, the elderly are not held to be morally superior but rather further along on a journey of life, with life seen as a movement of lifelong fulfillment whose consummation is found only in death and the afterlife. The sufferings of old age, in the traditional view, are seen against the wider background of the cosmos. The loss of that wider perspective is partly what deprives aging of meaning.[43]

However, that element of the cosmic scale concerns only the negative aspects of growing old, the inevitable losses and suffering. In many traditional societies these were balanced by a more positive view of human development. It is this positive movement that Philibert calls "the scale of ages." By this term, he means a ladder of lifespan development that individuals *may* ascend by way of spiritual perfectibility or growth in grace and wisdom, but such a spiritual ascent does not necessarily serve to protect the power position of the elderly. In some cases, the opposite may be true. In ancient Hindu doctrine, for example, the concept of *ashramas* (life stages of later life) would demand renunciation and the abandonment of power roles, not their protection.

Still less does spiritual development—movement up the scale of ages—serve to protect the individual against the downward movement, which includes the natural ravages of old age, such as the loss of loved ones and the prospect of our own death. Wisdom does not prevent suffering but allows us to find meaning in it. As the poet Louise Bogan said: finally, we want life to be understandable. However, if the traditional doctrine did not provide worldly compensations, it did promise a consolation for the inevitable losses. The modernist concept of the self contains no such potency toward spiritual fulfillment; it holds only an image of ceaseless flux and change, terminated by death.

The question of meaning in old age may signify at least two different things. First, does old age as a stage of life have a special meaning; and second, is old age that time of life when the full meaning of life itself is to be understood? These two different senses of meaning in old age cannot finally be separated. They coincide in the image of old age as a period of wisdom. We want the old to be wise, even if we are not certain anymore what wisdom might mean. Inasmuch as we ascribe wisdom to old age, then we expect those who are wise to understand the meaning of life.[44]

Since Aristotle, wisdom has been understood in a twofold sense: the knowledge that belongs to right action and the knowledge that belongs to right understanding.[45] If wisdom is seen as an ingredient of action, then the wise are those who have lived in harmony with life's true meaning or purpose: those who know how to live well. If wisdom is seen as an ingredient of understanding, then the wise are those with the widest and deepest insight into the nature of things. In both cases, whether wisdom belongs to contemplation or to action, we naturally look to the old ones for such wisdom. Wisdom is seen as an intellectual attribute, and the wise are those who, by definition, understand the meaning of life. Thus, the meaning of old age as a period of life seems intrinsically bound up with the larger question of whether life as such has a meaning and whether we can know that meaning.

However, today we feel that human beings can find no such objective meaning to the existence of human life in the cosmos. Existentialists may despair of this predicament, whereas secular humanists cheerfully propose projects to improve our condition. All are agreed on a basic premise: human life as such has no meaning other than the purposes that we give to it by our values and actions while we live. The traditional concept of the meaning of life is a delusion to be rejected. In the cosmic scheme, human existence is meaningless, and on the stage of history, every generation is on its own. We can learn nothing from the old. We give our lives whatever meaning they will have by the projects we choose.

This stance toward life, of course, deprives old age of any particular epistemological significance. We may owe the elderly decent treatment, but they can teach us nothing. Insofar as life in itself is felt to be meaningless, those who have lived longer will be no closer than younger people to an imaginary destination called "the meaning of life." Like the two characters in *Waiting for Godot,* we are in a position of merely accumulating more years in waiting, but the waiting brings us no closer to a goal that cannot exist. We live; we die; we are on our own, young and old alike.

The final paradox of the modernist ambivalence toward old age can be seen in the ideology and technology of medicine. Through biomedical technology, all possible effort is expended to keep elderly patients alive as long as possible. Thanks to the triumphs of public health and general affluence, an increasing proportion of the population now lives to experience the last stage of life—old age—which has itself been drained progressively of meaning. At the moment when the meaning of old age vanishes, we find that enormous economic resources are expended to prolong lives that have been deprived of any purpose. This paradox is apparent to anyone who works among the elderly in our society. As Jean-Pierre Dupuy notes, the contradiction informs our despair over the meaning of old age:

> The histories of pain, illness, and the image of death reveal that men have always managed to cope with these threats by giving them meaning, interpreting them in terms of what anthropologists call "culture." Today, however, the expansion of the medical establishment is closely connected with the spreading

> of the "myth" that the elimination of pain and the indefinite postponement of death are not only desirable objectives but that they also can be attained with the increased development of the medical system. A problematic arises: how do we give meaning to something that we seek to eliminate by all means? This issue is linked up with a more general characteristic of industrial—or postindustrial—societies, namely, that entire aspects of the human condition are becoming meaningless.[46]

Historians are divided on how far modernization in itself has led to a devaluation in the status or power-position of old age in comparison with other age groups.[47] What seems clear, nonetheless, is that modernization has led to a devaluation in the meaning of old age. Modernization has created distinctive social and cultural stresses on the position of old age in society, stresses that are experienced subjectively as a feeling that life in old age is meaningless. Among historically minded students of aging, there is recognition that the collapse of subjective meaning so often experienced in late life is a result of modern social policy; for example, the growth of retirement and the rise of bureaucratic systems in the human services.[48] There are some features of the culture of modernism, such as the diminishment of traditional community in favor of an isolated self, that have distinctly negative implications for old age. Then there are other features of modernism—preoccupation with time consciousness, for example—that are congenial to a recovery of a sense of the meaning of old age. In short, a central question to be examined is the connection between meaning and modernity with respect to human aging.

Modernization is not simply a matter of industrialization, technology, or changes in social institutions. It also signifies, in Peter Berger's words, a "modernization of consciousness."[49] The modern world reinforces a separation between the public and private world. When one belongs to the society of job holders, one participates in the public world, but upon retirement, the older person enters the invisible world of private existence. For the well-to-do, that private existence may be richly dedicated to pleasure or entertainment, but most elderly people find it difficult to anchor the self in a wider public world. In short, the dichotomies of public vs private and work vs leisure become transposed onto the life course itself.

The modernization of consciousness signifies a fundamental plasticity of the self—Lifton's "Protean Man"—infinitely open to change and transformation.[50] This Faustian image of the self is central to the project of modernism and its Promethean effort to overcome all limits.[51] The self is to be continually remade as new meanings are discovered or created. This process of perpetual remaking is what constitutes the modern project and the modern version of the meaning of life: to be always on the way, never to arrive. The significance for aging in the modern project is not merely its heroic or Promethean aspiration, which goes back to Leonardo, Michelangelo, and the spirit of the Renaissance. Heroic aspirations themselves have meaning only against a background of relatively stable tradition, but now tradition itself is in question. The cultural significance of the modern project lies in its conscious repudiation of tradition-as-limit.[52] There is no enduring set of traditions serving as a limit for the meaning that life can have. Consequently, old age, as the natural limit of life, is rejected along with those traditions that took such limits for granted.

Gerald Gruman has described the "modernization of the life-cycle."[53] The modernized life cycle is characterized by a twofold development, unprecedented in history: first, the separation of life into separate stages and age-groups; and second, the displacement of meaning into old age. It is the simultaneous convergence of these two trends that imperils the sense of meaning in late life. The shape of the life cycle in modern societies itself entails a displacement of the search for meaning into the last stage of life. This is a double displacement—from the afterlife into the last stage of life on earth and then from all earlier stages into the final stage. Since the cosmic sense of meaning has weakened, the image of fulfillment takes on the tone of secular ideologies. The most important of these is the displacement of free time and leisure into late life, supported by the spread of an ideology of retirement. The twofold development of the modern life cycle sets the stage for the contemporary problem of the meaning of old age.

One aspect of this development is the truncation of the life cycle. By truncation I mean division of the life course into the "three boxes of life," namely youth, adulthood, and old age, into which the social order allocates the three domains of education, work, and leisure.[54] The problem with this truncation of the life cycle is that it deprives us of any image of the unity of a human life, as Alasdair MacIntyre notes:

> Any contemporary attempt to envisage each human life as a whole, as a unity, whose character provides the virtues with an adequate *telos* encounters (an) obstacle. ...The social obstacle derives from the way in which modernity partitions each human life into a variety of segments, each with its own norms and modes of behavior. So work is divided from leisure, private life from public, the corporate from the personal. So both childhood and old age have been wrenched away from the rest of human life and made over into distinct realms. And all these separations have been achieved so that it is the distinctiveness of each and not the unity of the life of the individual who passes through those parts in terms of which we are taught to think and feel.[55]

The second aspect of the development of the modern life cycle is the dual displacement of meaning. By dual displacement, I mean first, the displacement of leisure/contemplation/meaning from the rest of adulthood into old age and second, the displacement of death/finitude/judgment from the afterlife into the present life. The modern distaste for contemplation is accompanied by a nostalgia for leisure, a wish to escape from haste, and a sentimental image of retirement as the "Golden Age" of life. Late life becomes the period when, freed from alienated labor, the "real self" can be fulfilled, as in the ideology of retirement. The modern world relentlessly projects meaning into the future. Thus work, savings, and deferred gratification are all strivings after a goal located in the future—in old age, but upon arriving at old age, there are many who would agree with Yeats's comment, "Life is a preparation for something that never happens."

Second, the dual displacement transposes the last judgment from eternity or afterlife into time. We see that displacement in the modern cult of death and dying, in "death education," in the submerged spirituality of life review as the final opportunity for meaning in life.[56] We are urged to "accept" death as a part of life, but we lack any cultural resources that might make acceptance possible.[57] Religious ideas of transcendence have not entirely disappeared. Instead, they have been transposed from the hope of salvation to the self-actualization promised by humanistic psychology, from the encounter with God to a new goal of freedom from anxiety. Medieval paintings once offered a deathbed scene with angels wrestling for the soul of the departing believer. Today, the dual displacement brings the last judgment into the nursing home and the intensive care unit. Under the dual displacement, the problem of meaning falls under the dominion of thanatology and gerontology, a new priesthood of professionals.

Both the truncation of the life cycle and the dual displacement of meaning are now reinforced by bureaucratic and professional institutions whose power and ideology are of great importance for the problem of meaning in old age. The institutions of the school, the workplace, and retirement have become bureaucratically ordained means of enforcing the "three boxes of life" for purposes of the political economy. The truncation of the life cycle is required by the economic system, because our current economy cannot provide employment for both the young and old. To get around that problem without challenging the basic structure of society, it is more feasible to disguise or displace free time under the names of "education" and "retirement." These phases of life then become longer and longer, as work is compressed into middle adulthood.

Unlike the situation in traditional societies, where myth and symbol shape the "rites of passage" over the life course,[58] the shape of the life course today is more and more subject to professional expertise of specialists in this or that segment of the life course. Just as scientific management of human beings is needed to provide a smooth flow of labor in the assembly line, so a smooth flow of age groups is maintained by professional management for the different stages of life. The truncation of life stages is ordained not by nature, but by the planning and control needs of the larger society. There is, accordingly, a need to manage the periodic crises of different life stages: adolescence, midlife crisis, preretirement planning, and so on.

Unlike traditional laissez-faire capitalism, where competition is theoretically open-ended, late capitalism is monopoly capitalism in which the state plays a critical role in organizing labor markets, planning research and development, and organizing the social infrastructure, including the system of education, old age services, and, in general, the broad spectrum of social welfare activities required for existence in society.[59] Consequently, neither education nor retirement can be left to private decision-making or discretionary choice: both the first and the last stages of the life course are absorbed into the planning and control systems of society. Decisions now fall under the direction of specialized professions—child development, gerontology, adult counseling, industrial psychology, and so on—which in turn provide normative guidance for organizing the life course from birth until death. The rationale for management is provided by a covert ideology of normative life stages that is itself subject to changing interpretations.

The heart of the problem for management is clear. There is an ideological contradiction in the fact that the society is increasingly driven by two opposing demands: a cultural drive for maximum autonomy on the one hand, and an economic drive for efficiency and control on the other.[60] The progressive truncation of the life course promotes efficiency but contradicts individual autonomy. At points of life course transition or crisis, contradictions are exposed. It is the task of the ideology of lifespan development to mask these contradictions at every point.

The newest of these covert ideological interpretations is the good news from gerontologists that we are about to enter an "age-irrelevant" society.[61] Nothing could be further from the truth, but the proclamation of an end to the tyranny of normative life stages will be welcomed in a society where the denial of aging and death is still deeply rooted: "You're only as old as you feel." New interpretations of lifespan development end up proclaiming the American bootstrap ideology of self-determination ("Pull your own strings," "You're never too old to learn," and so on). However, the distribution of opportunities for lifespan development is bound to forces of history, culture, and political economy. This contradictory relationship between political economy and the ideology of the life course stands unexamined by the professional ideologies of life planning and lifespan development. On the contrary, images of the life course furnished by the social sciences are held to be a form of "value-free" knowledge independent of human interests.[62]

The covert ideologies of lifespan development present themselves as apolitical, "scientific" conclusions devoid of value-commitments. Yet they inevitably share the dominant liberal ideology: improvement of society through applied scientific knowledge, which is eventually to lead to greater individual self-determination. In the ideology of gerontology, this liberal individualism is expressed in a refusal to accept age categories as a basis for social decisions (such as opposition to mandatory retirement). This refusal of the truncation of the life course is appealing, but the hope of breaking up the three boxes of life remains utopian as long as the relation between culture, ideology, and political economy stands unexamined. The liberal hope of maximizing individual opportunity in opposition to restrictive categories, including age, is another of the cultural contradictions of capitalism.[63]

Even a weakening of rigid truncation may not necessarily lead to a favorable outcome. As old boundaries are erased, new forms of domination take their place. An age-irrelevant society is indeed taking shape, but not in the utopian form imagined. Both young and old are affected today. The rise of the "young old" is matched by the "disappearance of childhood."[64] Under the impact of television, the idealized innocence of childhood is becoming a thing of the past. Both the innocence of childhood and the wisdom of age are relics of a premodern world. For young and old alike, sophistication and activity become the dominant style. In both cases, the early and late stages of the life course are absorbed into the endless present of perpetual young adulthood, now the dominant ideological image of the ideal worker-consumer.

Here the modern project of an existence without boundary—the perpetual remaking of the self—comes full circle. In Auden's phrase, "we are all contemporaries," but only because history, including life history, has been abolished. We live in an information economy, a world where instantaneous media imagery juxtaposes past and future, fantasy and reality. In this world, it becomes easy to believe that the self is only a metaphor, a choice to be discarded or refashioned at will. It is a worldwide Disneyland where the world of tomorrow and the antique past are equally available and equally unreal.

These remarks are critical of the covert ideology of lifespan development, but I do not conclude that scientific inquiry into lifespan development psychology lacks importance for a deepened sense of meaning in old age. On the contrary, the idea of lifespan development psychology is itself the great and indispensable myth of our time. For better or for worse, it remains our hope for some form of meaning embracing the whole of life. Further, the emergence of a new image of human development over the entire lifespan has the greatest importance today in providing an underpinning for movements of social change that could improve the quality of life for older people. Whether as ideology, myth, or science, we need some image of the whole of life.

However, where will the cultural resources for this new image be found? I suggest that the future of lifespan development psychology may depend very much on concepts drawn from the humanities rather than on the biological models that have prevailed in the past.[65] Such a paradigm of lifespan development, we would argue, would be hermeneutic: that is, it would presuppose that acts of interpretation—specifically, self-interpretation—lie at the center of the developmental drama.[66] We would move, in short, "from system to story,"[67] away from cybernetic models toward a fully historical understanding of the stories by which we live out our lives.[68]

We would begin to see that there is no such thing as the "natural" human life course. The scientific project to discover such a natural pattern is misconceived and must end in mystification. Even the empirical discovery of uniform or fixed periods of stability and change—as Daniel Levinson and his associates report from clinical data—would become a problem of hermeneutical, not biochronological, significance.[69] Why, for instance, does it turn out to be the case that critical life transitions—the so-called "midlife crisis," for example—are linked to round-number age-transitions—age thirty or age forty? There is no question that these age transitions are often important "marker events" in the psychological development of adults, as Levinson shows, but Levinson does not even offer an explanation for the numerical regularities he finds in these transitions.

An explanation for these periodicities will not be found in the model of the Periodic Table of Elements. The true explanation is to be found in the symbolic meaning of age numbers—the "developmental numerology" of our developmentally obsessed culture. An historical analogy makes the point clear. As the millennial year 1000 approached, all of medieval Christendom waited with fear and trembling for an apocalyptic end to the world. We may fully expect a comparable outpouring of mythic energy in our own time when the millennial year 2000 comes round—less than a decade away, but, obviously, no natural law dictates the upsurge of such collective feelings. Rather, these round numbers galvanize our collective imagination around dramatic temporal symbols. In the same way, approaching age 30, 40, or 65 can provoke anxieties of transition, but this developmental numerology demands a hermeneutical, not a positivistic, explanation. It opens up questions of *meaning*.

It is preeminently to the disciplines of the humanities that we turn for illumination of questions of meaning. The primary disciplines of the humanities—philosophy, history, and literature—are concerned with the "three Cs": criticism, continuity, and communication.[70] Critical questioning (philosophy), the continuity of time (history), and the communication of shared meanings (literature) are all central in understanding the problem of meaning of old age. Analytic philosophy has approached questions of the meaning of life from a logical, atemporal standpoint, whereas literary criticism has incorporated the temporal depth of meaning into the study of autobiography. We need an approach to the hermeneutics of aging that can do justice to the time-bound nature of aging as well as to the timeless philosophical questions involved. I do not want to imply that the hermeneutic approach alone is adequate. Meanings and acts of interpretation are never purely subjective transactions occurring outside culture and history.[71] They are part of this larger context.

In recent years, we have seen a rapprochement between lifespan development psychology and concern for spiritual growth.[72] An encouraging development has been the appearance of work that draws on the resources of spirituality.[73] The spiritual traditions have never accepted the idea that human fulfillment is the product of social roles or relentless activity in the world, nor have they accepted the idea that the meaning of old age can be separated from the meaning of life as a whole. As I argued earlier, I believe that no assessment of the search for meaning in late life will be adequate without a reappraisal of the role of activity and contemplation in the modern world. There may well be gifts of understanding that are, in an existential sense, reserved for the last stage of life. In casting aside the values of contemplation from our benevolent desire to "do something" for the aged, we may have missed what is of still greater importance—namely, what the aged can give to us: a reminder, perhaps, of the finality of life, which could be a precious gift for those who can receive it.

Notes and References

[1]M. C. Nelson, ed. (1977) *The Narcissistic Condition: A Fact of Our Lives and Times.* Human Sciences Press, New York, NY.

[2]P. Rieff (1966) *The Triumph of the Therapeutic.* Harper and Row, New York, NY; C. Lasch (1978) *The Culture of Narcissism.* W. W. Norton, New York, NY.

[3]L. Wittgenstein (1961) *Tractatus Logico-Philosophicau* (D. F. Pears and B. F. McGuinness, eds.), Routledge and Kegan Paul, London, UK, pp. 149–151.

[4]E. Jaques (1965) Death and mid-life crisis. *International Journal of Psycho-Analysis* **46 (4),** 502–512.

[5]*See* K. Boulding (1956) *The Image.* University of Michigan Press, Ann Arbor, MI. Boulding is heavily influenced by F. Pollak, *The Image of the Future.*

[6]M. Munitz (1966) *The Mystery of Existence.* Appleton-Century-Crofts, New York, NY.

[7]*See* S. Korner (1955) *Kant.* Penguin Books, New York, NY. For more detail, *see* A. C. Ewing (1970) *A Short Commentary on Kant's Critique of Pure Reason.* University of Chicago Press, Chicago, IL.

[8]E. Erikson (1968) The human life cycle. *International Encyclopedia of the Social Sciences.* For an appraisal, *see* P. Roazen (1977) *Erik Erikson: The Limits of a Vision.* Free Press, New York, NY.

[9]D. Ross (1964) *Aristotle.* Barnes and Noble, New York, NY; and J. H. Randall, Jr. (1960) *Aristotle.* Columbia University Press, New York, NY. A reconstruction of the Aristotelian point of view applied to lifespan development can be found in D. Norton (1976) *Personal Destinies.* Princeton University Press, Princeton, NJ. Norton is one of the few American philosophers who has approached the concept of life stages as a significant philosophical problem.

[10]J. M. Rist (1969) *The Stoic Philosophy.* Cambridge University Press, New York, NY.

[11]On the question of "rational suicide" based on declining quality of life in old age, *see* H. R. Moody (1983) Can suicide on grounds of old age be ethically justified? in *Thanatology and Aging* (M. Tallmer, ed.), Columbia University Press, New York, NY.

[12]*See* E. Erikson (1950) *Childhood and Society.* W. W. Norton, New York, NY, and (1982) *The Life Cycle Completed: A Review.* W. W. Norton, New York, NY.

[13]R. C. Peck (1968) Psychological development in the second half of life, in *Middle Age and Aging* (B. Neugarten, ed.), University of Chicago Press, Chicago, IL.

[14]R. J. Lifton (1979) *The Broken Connection: On Death and the Continuity of Life*. Simon and Schuster, New York, NY.

[15]C. Tart (1977) *Transpersonal Psychologies*. Harper and Row, New York, NY.

[16]R. Otto (1923) *The Idea of the Holy* (J. W. Harvey, trans.), Oxford University Press, New York, NY. On contemplation, *see* F. Schuon (1975) *The Transcendent Unity of Religions* (P. Townsend, trans.), Harper and Row, New York, NY.

[17]R. Panikkar (1981) The contemplative mood: A challenge to modernity. *Cross Currents* **31,** 261–271.

[18]S. de Beauvoir (1972) *The Coming of Age* (P. O'Brien, trans.), G. P. Putnam's Sons, New York, NY.

[19]T. R. Cole (1983) The "enlightened" view of aging: Victorian morality in a new key. *Hastings Center Report* **13,** 34–40.

[20]On the hegemony of the *vita activa* in the modern age, *see* H. Arendt (1958) *The Human Condition*. University of Chicago Press, Chicago, IL.

[21]F. Schuon (1972) *Understanding Islam*. Penguin Books, New York, NY, p. 35.

[22]K. Jaspers (1951) *Man in the Modern Age*. Routledge and Kegan Paul, London, UK.

[23]K. Britton (1969) *Philosophy and the Meaning of Life*. Cambridge University Press, New York, NY; W. D. Joske (1974) Philosophy and the meaning of life. *Australasian Journal of Philosophy* **52 (2),** 93–104. Reprinted in E. D. Klemke, ed. (1981) *The Meaning of Life*. Harvard University Press, Cambridge, MA.

[24]R. Nozick (1981) *Philosophical Explanations*. Cambridge University Press, New York, NY.

[25]J. Needleman (1968) Why philosophy is easy. *Review of Metaphysics* **22 (1),** 3–14. More recently, Needleman (1982) *The Heart of Philosophy*. Alfred A. Knopf, New York, NY.

[26]V. Frankl (1963) *Man's Search for Meaning*. Beacon Press, Boston, MA.

[27]R. N. Butler (1963) The life review: An interpretation of reminiscence in the aged. *Psychiatry* **26,** 65–76. Successful aging and the role of the life review, in S. H. Zarit (1977) *Readings in Aging and Death*. Harper and Row, New York, NY. For a bibliography, *see* H. R. Moody (1984) Bibliography on life-review, in *The Uses of Reminiscence* (M. Kaminsky, ed.), Haworth Press, New York, NY.

[28]S. Merriam (1980) The concept and function of reminiscence: A review of the research. *The Gerontologist* **20,** 604–609.

[29]K. Baier, The meaning of life, in *The Meaning of Life* (Klemke, ed.), pp. 81–117.

[30]Ibid., p. 115.

[31]*See* D. M. Frame (1955) *Montaigne's Discovery of Man.* Columbia University Press, New York, NY.

[32]*See* L. Wittgenstein (1953) *Philosophical Investigations* (G. E. M. Anscombe, trans.), Oxford University Press, New York, NY. This is the source for the important idea of family resemblance as the key to the unity of concepts within a language game. Cf. R. Bambrough (1960,1961) Universals and family resemblances. *Proceedings of the Aristotelian Society* **61.**

[33]S. Pepper (1942) *World Hypotheses.* University of California Press, Berkeley, CA; and T. Kuhn (1970) *The Structure of Scientific Revolutions.* University of Chicago Press, Chicago, IL.

[34]J. Olney (1973) *Metaphors of Self.* Princeton University Press, Princeton, NJ.

[35]L. Trilling (1980) *Sincerity and Authenticity.* Harcourt Brace Jovanovich, San Diego, CA.

[36]J. Olney (1980) *Autobiography: Essays Theoretical and Critical.* Princeton University Press, Princeton, NJ. *See* Olney's introductory essay for an account of what appears to be the *cul-de-sac* of the contemporary autobiographical consciousness, where linguistic acts abolish both self and history.

[37]J. Varner Gunn (1982) *Autobiography: Toward a Poetics of Experience.* University of Pennsylvania Press, Philadelphia, PA.

[38]The examples of Augustine, Montaigne, Rousseau, and Kierkegaard are instances of philosophical thought that is at the same time personal and autobiographical. However, in the history of philosophy, it is possible to identify instances of what might be called pseudo-autobiography, such as Descartes' *Discourse on Method.* In our time, philosophers, such as Bertrand Russell, A. J. Ayer, or George Santayana, can write elegant autobiographies that seem to have no explicit relation to their thought—another instance of the gulf between public and private worlds. The relation between the literary style and the conceptual intentions of philosophers remains an unwritten chapter in the history of ideas.

[39]Cf. Mircea Eliade (1959) *Cosmos and History: The Myth of the Eternal Return.* Harper and Row, New York, NY.

[40]P. Berger (1973) *The Homeless Mind: Modernization and Consciousness.* Random House, New York, NY.

[41]M. Philibert (1968) *L'Echelle des Ages.* Editions du Seuil, Paris; (1974) The phenomenological approach to images of aging. *Soundings* **57,** 33–49.

[42]D. Stannard (1977) *The Puritan Way of Death.* Oxford University Press, New York, NY.

> [43]We are witnessing now an attempt to eliminate the darker, more painful aspects of human living, no longer by rising above them (and thereby gaining in stature), but either by abolishing them—which is impossible since they lie in the nature of things—or by pretending they do not exist. It was possible for the men of other times to accept these conditions because life as such was situated in an infinitely wider context. They knew that however deeply involved they might be in the scenario of suffering and loss, they were not by nature totally submerged in it. Experience taught them to look elsewhere for peace and for perfection, and faith assured them that there are indeed other dimensions than those that seem to hem us in. Today, most people are confined in a place that knows no "elsewhere," trapped with wild beasts that tear their flesh and from which they cannot escape. It is hardly surprising that they need to be drugged to be able to exist in such a situation.

G. Eaton (1977) *King of the Castle: Choice and Responsibility in the Modern World.* The Bodley Head, London, UK., p. 37.

[44]*See* J.-R. Staude (1981) *Wisdom and Age.* Ross Books, Berkeley, CA; and V. Clayton and J. Birren (1980) The development of wisdom across the lifespan: A reexamination of an ancient topic, in P. Baltes and O. Brim, *Lifespan Development and Behavior.* Academic Press, New York, NY.

[45]On the concept of wisdom in the classical and medieval tradition, *see* J. D. Collins (1962) *The Lure of Wisdom.* Marquette University Press, Milwaukee, WI.

[46]J.-P. Dupuy (1980) Myths of the information society, in *The Myths of Information: Technology and Postindustrial Culture* (K. Woodward, ed.), University of Wisconsin Press, Madison, WI, p. 9.

[47]D. Hackett Fischer (1978) *Growing Old in America.* Oxford University Press, New York, NY, pp. 232–269; and W. A. Achenbaum and P. Stearns (1978) Old age and modernization. *The Gerontologist* **18,** 307–312.

[48]W. Graebner (1980) *A History of Retirement: Meaning and Function of an American Institution, 1885–1978.* Yale University Press, New Haven, CT; P. Townsend (1981) The structured dependency of the elderly: Creation of social policy in the twentieth century. *Aging and Society* **1 (1),** 5–28.

[49]P. Berger (1973) *The Homeless Mind: Modernization and Consciousness.* Random House, New York, NY.

[50]R. J. Lifton (1970) Protean man, in his *History and Human Survival.* Random House, New York, NY.

[51]M. Berman (1982) *All That Is Solid Melts into Air: The Experience of Modernity.* Simon and Schuster, New York, NY. Berman is helpful in distinguishing between *modernization* as a politico-economic transformation and *modernity* as a cultural condition. This distinction is ignored in the gerontological literature, resulting in a confusion of the status and meaning of old age.

[52]E. Shils (1981) *Tradition.* University of Chicago Press, Chicago, IL; and H. Smith (1976) *Forgotten Truth: The Primordial Tradition.* Harper and Row, New York, NY.

[53]G. J. Gruman (1978) Cultural origins of present-day "age-ism": The modernization of the life cycle, in *Aging and the Elderly: Humanistic Perspectives in Gerontology* (S. F. Spicker, K. M. Woodward, and D. Van Tassell, eds.), Humanities Press, Atlantic Highlands, NJ.

[54]R. Bolles (1979) *The Three Boxes of Life.* Ten Speed Press, Berkeley, CA. *See also* F. Best (1980) *Flexible Life Scheduling: Breaking the Education-Work-Retirement Lockstep.* Praeger Publishers, New York, NY.

[55]A. MacIntyre (1981) *After Virtue.* University of Notre Dame Press, Notre Dame, IN, p. 190.

[56]P. Aries (1974) *Western Attitudes Toward Death* (P. M. Ranum, trans.), Johns Hopkins University Press, Baltimore, MD; and J. Choron (1963) *Death and Western Thought.* Collier Books, New York, NY.

[57]E. Becker (1973) *The Denial of Death.* Free Press, New York, NY; and J. Choron (1964) *Death and Modern Man.* Collier Books, New York, NY.

[58]Compare S. Moore and B. Myerhoff, eds. (1977) *Secular Ritual.* Van Gocum, Amsterdam; and V. Turner (1969) *The Ritual Process: Structure and Anti-Structure.* Routledge and Kegan Paul, London, UK.

[59]J. O'Conner (1970) *The Fiscal Crisis of the State.* St. Martin's Press, New York, NY.

[60]D. Bell (1973) *The Coming of Post-Industrial Society.* Basic Books, New York, NY. Bell argues that the economy and the culture of postindustrial society pull in diametrically opposite directions: the economy demands increased specialization and limits, whereas the culture promises indefinite augmentation of the self.

[61]B. Neugarten (1974) Age groups in American society and the rise of the young-old. *Annals of the American Academy of Political and Social Science* **320,** 187–198; and (1983) *Age or Need?* Sage Publications, Beverly Hills, CA.

[62]J. Habermas (1975) *Knowledge and Human Interests* and *Legitimation Crisis.* Beacon Press, Boston, MA. For an introduction, *see* T. McCarthy (1978) *The Critical Theory of Jurgen Habermas.* MIT Press, Cambridge, MA.

[63]D. Bell (1976) *The Cultural Contradictions of Capitalism*. Basic Books, New York, NY.

[64]N. Postman (1982) *The Disappearance of Childhood*. Delacorte, New York, NY; and M. Winn (1983) *Children Without Childhood*. Pantheon, New York, NY.

[65]*See* K. F. Riegel (1972) The influence of economic and political ideologies upon the development of developmental psychology. *Psychological Bulletin* **78,** 29–141; On the history of psychological gerontology, in *The Psychology of Adult Development and Aging* (C. Eisdorfer and M. Powell Lawton, eds.), American Psychological Association, Washington, DC, 1973; and Toward a dialectical theory of development. *American Psychologist* **31,** 689–700.

[66]J. Bleicher (1980) *Contemporary Hermeneutics*. Routledge and Kegan Paul, London, UK; and L. C. Watson (1976) Understanding a life history as a subjective document: Hermeneutical and phenomenological perspectives. *Ethos* **4,** 95–131.

[67]D. Burrell and S. Hauerwas (1977) From system to story: An alternative pattern for rationality in ethics, in *Knowledge, Ethics, and Belief* (H. Tristram Englehardt and D. Callahan, eds.), Hastings Center, Hastings on Hudson, NY, pp. 111–152.

[68]*See* C. G. Prado (1983) Aging and narrative. *International Journal of Applied Philosophy* **1 (3),** 1–14. What MacIntyre had termed the "narrative unity of a human life" constitutes, in essence, a regulative ideal for the meaning of old age. Yet the notion of narrative unity itself may not be consistent with the cultural mood of modernism.

> One feature that links the [modernist] movements...is that they tend to see history or human life not as a sequence or history, not as an evolving logic. ...Modernist works frequently tend to be ordered, then, not on the sequence of historical time or the evolving sequence of character, from history or story as in realism or naturalism; they tend to work spatially or through layers of consciousness, working toward a logic or metaphor or form.

M. Bradbury and J. McFarlane (1976) *Modernism*. Penguin Books, New York, NY, p. 50. The devaluation of character, sequence, and narrative has its counterpart in the destruction of ideals of virtue, wisdom, and meaning: all ingredients of the narrative unity of a human life.

[69]D. Levinson (1978) *The Seasons of a Man's Life*. Alfred A. Knopf, New York, NY.

[70]A. W. Levi (1970) *The Humanities Today*. Indiana University Press, Bloomington, IN. On the humanities and gerontology, *see* D. Van Tassell, ed.

(1979) *Aging, Death and the Completion of Being*. University of Pennsylvania Press, Philadelphia, PA.

[71]*See* P. Rabinow and W. M. Sullivan, eds. (1979) *Interpretive Social Science,* University of California Press, Berkeley, CA.

[72]J. W. Fowler (1981) *Stages of Faith: The Psychology of Human Development and the Quest for Meaning.* Harper and Row, New York, NY.

[73]H. Nouwen (1974) *Aging: The Fulfillment of Life*. Doubleday, Garden City, NY; and E. C. Bianchi (1982) *Aging as Spiritual Journey*. Crossroads, Los Angeles, CA.

Oedipus and the Meaning of Aging

Personal Reflections and Historical Perspectives

Thomas R. Cole

"Old age," someone once said, "is like a rock on which many founder and some find shelter." When I was a child, the old people in my family were a sheltering rock; they occupied a place in my inner landscape that has only recently begun to change. My grandparents and great aunts always seemed the same. They were there every weekend, on holidays, whenever we needed them. My brother, sister, and I ate at their tables, roamed their houses, climbed in their yards, ravished their presents, and took for granted their immortality.

When my father died at the age of 27, I was immediately transformed into an aged four-year-old, a *senex puer.* For many years, I nursed private intuitions of ultimate truth alongside feelings of loneliness, guilt, and depression. My father's death broke the sequence of generations; I felt that I too would die young, and could not live childhood's normal innocence and exuberance.

My grandfathers died when I was nine and 16. Irving—the quiet, cigar-smoking, blueprint shop entrepreneur who cried every day of his life after the death of his son—disinherited us when we were legally adopted by my mother's second husband. The night Irving died, I had been on a local television show with my Cub

Scout pack. My mother said he must have died happily after seeing me on TV. I wondered more darkly about the connection. Jack, the hard-driving, self-made son of a Jewish immigrant tailor was a Yale graduate whose intensity, ambition, and success both inspired and haunted me. I felt great pride at being a pallbearer at his funeral, where my uncontrollable crying disturbed an otherwise dignified affair.

The deaths of my grandfathers were painful, but they did not undermine the sense of continuity derived from my grandmothers, who seemed to go on without change. Until my late 30s, they maintained their independence, each somewhat stern and difficult in her own way, each fiercely loyal and proud. Their existences helped frame my own; they demanded little and gave much.

Then, in the spring of 1986, after being forced out of the blueprint business she had run for 50 years, Reba (Irving's wife and my father's mother) began to lose the ability to direct her own affairs. Failing vision from cataracts, acute glaucoma, severe arthritis, stomach trouble, and finally Alzheimer's disease broke her. Yet she fought furiously against me when I arranged a conservatorship and round-the-clock care for her. Both she and my one-year-old daughter were in diapers; neither could walk more than a few steps without falling; neither lived in the adult world of chronologically measured secular time.

In the spring of 1987, my mother's mother entered the hospital for the first time since her second daughter was born. She died two weeks later. Helen had carried tremendous weight in the family; she possessed an unwavering sense of dignity and high standards, and she also possessed the wealth accumulated by Jack. She had always seemed invulnerable, as if she might actually outlive death, if not the rest of us. Her death was shocking despite her 87 years; it felt like the rope of a great anchor had suddenly come unraveled, leaving a sense of turbulence and weightlessness. Six weeks later, my grandfather's sister died and was buried in the same family plot that held my grandparents, their parents, and my father.

The "rock" of old age no longer feels so secure: My mother and stepfather have begun to negotiate its coastline, whose waters seem stormier and closer than before, but if the rock of old age seems more perilous, it is also more enchanting. Amidst waves of brokenness, rage, and loneliness, I sense its possibilities for wholeness and self-transcendence. Perhaps, if I am blessed, I may someday become a *puer senex,* an innocent and playful old man.

During the last decade, these feelings have fueled my inquiry into aging. Stirred by the search for personal meaning, I have been exploring the topography of old age, charting its cultural forms. I am an historian whose questions are existential and moral: Why do we grow old? Does aging have an intrinsic purpose? Is old age the culmination or the dreary denouement of life's drama? Is there anything important to be done after children are raised and careers completed? Are there perduring "gifts reserved for age"? Has death always cast its shadow over old age? What are the rights and responsibilities of older people? What are the virtues of old age? Has there ever really been a "good" old age?

Despite the rapid aging of Western populations since the nineteenth century and the vast gerontological literature that has appeared since World War II, such questions have received remarkably little attention.[1] Our culture is not much interested in *why* we grow old, or how we *ought* to grow old, or what it *means* to grow old. Modern biomedical and social science, the dominant voice in our public discourse about aging, is primarily interested in *how* we age in order to understand and control the aging *process.* Like other aspects of our biological and social existence, aging has been brought under the dominion of scientific management,[2] which hopes the primary questions will soon be "whether to wither and why."[3]

The rise of modern gerontology and geriatrics has certainly elevated the physical health of most older people. Hearing aids, pacemakers, hip replacements, and cataract surgery, to name only a few medical innovations, contribute enormously to the quality of late life, but for all its practical accomplishments, the intellectual hegemony of science contributes to the cultural and symbolic impoverishment that has beset the last half of life since the late nineteenth century. Gerontology and geriatrics encourage the perception of aging as a technical problem faced only by old people. Focusing narrowly on a reified "problem of old age," apart from the actual lives and cultural representations of people growing older, the scientific management of aging denies our universal participation and solidarity in this most human experience. However, *homo sapiens* are spiritual animals, and we need love and meaning no less than food, clothing, shelter, and health care.

For at least the last 60 years, Western observers have sensed the impoverishment of meaning in later life. In 1922, G. Stanley Hall noted that modern progress both lengthened old age and drained it of substance.[4] During the 1930s, Carl Jung observed that many of his patients found little meaning or purpose in life as they grew older.[5] In 1949, A. L. Vischer wondered whether "there is any sense, any vital meaning in old age."[6] Fifteen years later, Erik Erikson argued that, lacking a culturally viable ideal of old age, "our civilization does not really harbor a concept of the whole of life."[7] Summarizing a large volume of research in 1974, Irving Rosow claimed that American culture provided old people with "no meaningful norms by which to live."[8] More recently, Leopold Rosenmayr claimed that the position of the elderly in Western society "can only be reoriented and changed if viable ideals, 'existential paradigms,' become visible and receive some social support."[9]

The meaning of old age in any historical era is inevitably linked to a culture's understanding of life's meaning, and understanding life's meaning, as Harry R. Moody argues, finally involves some intuitive grasp of its wholeness or unity.[10] Until recently, Western culture has relied heavily on two archetypal themes to represent such intuitions: the division of life into ages (or stages) and the metaphor of life as a journey. Classical antiquity first connected these themes and wove them into beliefs about the nature of human existence and the cosmos. In the Middle Ages, Christian writers adopted Greco-Roman ideas about the "ages of life" and transformed the "journey of life" into a sacred pilgrimage. From the Reformation, which forged the modern iconography of the life cycle, until World War I, these images provided an existentially vital and culturally powerful framework that helped sustain the social meaning of aging and old age.

The archetypal power of the ages and journey motifs derives from their capacity to help us think about the mystery of human temporality. Virtually universal, they appear in many cultures, not as cognitive abstractions, but as ritualized elements of individual and social life.[11] Each offers a way of conceiving a fragmented, sometimes chaotic, ever-changing "life time" as a unified whole. This imagined whole, within which various parts of life can be located, guarantees a coherent (if not always happy) place for aging and old age.

By the fifth century B.C.E., Greek legend and custom had divided human life into three ages, each corresponding to a generation, each possessing its own set of natural characteristics and prescribed behavior.[12] Aristotle formalized this threefold division in the *Rhetoric,* where he discussed the ages of growth, stasis, and decline.[13] Hippocrates described the four physiologically determined ages—the most common scheme until the late Middle Ages, when the astrologically based system of seven ages was translated into the vernacular and eventually immortalized by Shakespeare's cynical Jaques.

In *De Senectute,* Cicero identified the philosophical bedrock beneath the ages of life doctrine: the belief that, despite the diversity of size, appearance, ability, and behavior that characterizes the different ages, the human lifespan nevertheless constitutes a single natural order. "Life's racecourse is fixed," wrote Cicero, "Nature has only a single path and that path is run but once, and to each stage of existence has been allotted its appropriate quality."[14]

The ages of life, then, offer a broadly unified view of a human lifetime; the processes of birth, growth, maturity, decay, and death appear as parts of the cycle of organic life, but as Abraham Heschel reminds us, being human is not identical with human being. "The passage of being..." he writes, "is marked and fixed: from birth to death. The passage of being human leads through a maze: the dark and intricate maze we call the inner life...That maze must not be conceived as a structure...it is an exuberance that goes on, frequently defying pattern, rule, and form. The inner life is a state of constantly increasing, indefinitely spreading complexity."[15]

Growing up and old, therefore, is not only an objective process, rooted in our biological existence, structured by social and historical circumstances. It is also a personal experience, an incalculable series of events, moments, and acts lived by an individual person. This experience, this passage through the maze of the inner life, has been traditionally represented by the journey of life—a metaphor that operates by narrating diverse experiences in time and space, bringing them under control of a single unifying purpose.

The journey is among the most pervasive themes in world literature. Monica Furlong writes:

> Wherever we look in folk-tale, art, religious teaching, we find evidence...of a hero setting out from a haven which has become constricting, undergoing a series of adventures which have a transforming effect, and eventually reaching, or failing to reach, his goal, or prize, or spiritual home...The journey is essentially a learning process, often of an acutely painful and humiliating kind, so that bit by bit, the hero becomes a changed man whose needs and satisfactions may be very different from those of the callow youth who set out.[16]

The stories of Gilgamesh, Job, Odysseus, or Aeneas reflect the power of this metaphor in antiquity, when poets first articulated the two major shapes of the journey: circular progression toward the renewal or restoration of the traveler, achieved through a homecoming; and linear progression from a situation of social or intellectual disorder to one of order.[17] When the oral tradition of the Jews was written down in the *Pierkei Avot (Wisdom of the Fathers),* the journey of life figured prominently in the teaching of Rabbi Akaviah ben Mahalalel: "Ponder three things and you will avoid falling into sin: Know your origin, your destination, and before Whom you will be required to give an accounting." The Christian tradition, of course, has used the theme to convey the spiritual life "strangers and pilgrims" (Hebrews XI:13) in search of God.

Although both themes offer images of life as a whole, each leaves out the other's essential insight. The ages of life envision the externally viewed, socially and biologically staged process of the human life cycle. The journey of life, on the other hand, foregrounds the fluid and unique qualities of individual experience, the spiritual drama of the traveler's search. The ages (or stages) of life offer a compelling framework to help answer the question "What is Man?" The journey of life frames an irreducibly narrative and personal response to the question "Who is Man?"[18]

Since the human animal is simultaneously biological and spiritual, social and individual, existential and historical, a deeper understanding of human "life-time" requires some imaginative juxtaposition of these themes—in which neither is reduced to the other and both create a whole greater than the sum of their parts. In *Oedipus Rex* and *Oedipus at Colonus,* I will argue in the remainder of this chapter, Sophocles achieved such a dynamic vision, and thereby generated profound insights into aging, death, and generational succession that bear significantly on our aging society in the late twentieth century.

It is important to remember, however, that the value of subjectivity, and the moral and spiritual importance of each individual are modern ideas. Since antiquity placed virtually no value on the experience of ordinary individuals, ancient writers used the journey of life to characterize extraordinary individuals, or heroes. To be an individual at the height of classical Athenian culture, for example, meant to be a person whose special destiny set him apart from the secure values of the community. Only a hero, and a masculine one at that, was thought to possess the power and intelligence to endure the isolation, pain, and uncertainty of the long journey to fulfillment. It is appropriate, therefore, that Sophocles wove these themes together to depict the tragic figure of Oedipus.

Like Freud, Sophocles used the legend of Oedipus to emphasize the sad inevitability of generational conflict—conflict guaranteed by the tensions that arise because each generation is destined to rise, decline, die, and be replaced by a new generation. However, whereas Freud pointed an accusing finger at the young Oedipus, recent commentators have pointed out that the tragedy is set in motion by his father, King Laius.[19] Laius attempts to evade the oracle's prophecy (that he will be killed by his own son) by leaving the infant Oedipus on a hillside to die, but a compassionate shepherd takes Oedipus to Corinth, where he is raised by a couple he believes are his real parents. Laius' actions, which signify the attempt to deny mortality and generational succession, only seal his fate, which he receives fifteen years later at the hands of the unknowing Oedipus.

Oedipus, we will remember, did not ascend to the throne of Thebes through the murder of his father, King Laius. Before the action of *Oedipus Rex* begins, he saves the city from the monster Sphinx, who is strangling its inhabitants for being unable to answer her riddle: "What goes on four legs in the morning, two legs at noon, and three legs in the afternoon?" "Man," replies Oedipus; he crawls as an infant, walks upright in his prime, and hobbles with a cane in old age. A human is the animal whose three ages of life demand different forms of movement, but what are the origins and destiny of this movement?

Here, Sophocles is nudging us toward a deeper understanding of the riddle: we come to know the ages of life by living a personal journey from vulnerable, dependent childhood to apparently independent adulthood and on to vulnerability and dependence again in old age. Moving from stage to stage, we come to know ourselves. For all its apparent brilliance, Oedipus' solution to the Sphinx's riddle cannot ultimately save him or Thebes. Unrelated to his own origins or destiny, his answer is incomplete, uninformed by self-knowledge. Oedipus remains a riddle to himself.

In identifying humans as the creatures whose lives resemble the rising and setting of the daily sun, Oedipus seems to affirm the ineradicable limits of the human condition and the necessity of generational continuity. Yet the great problem solver must first set out on the tragic road to self-knowledge before he will fully understand the riddle of human existence in time.

Years after the death of Laius, when the action of *Oedipus Rex* begins, the city of Thebes is suffering from a plague sent by the gods for the unsolved murder of the old King. With the same intellectual self-confidence he displayed in solving the Sphinx's riddle, Oedipus proclaims that he will find the murderer and banish him from Thebes. Thus slowly unravels the mystery of Oedipus' own origins, the story of patricide and incest leading to blindness, horror, and exile.

For most of us, I suspect, this is where the story of Oedipus ends, but for Sophocles, who completed *Oedipus at Colonus* at age 89, Oedipus' self-blinding and exile are the beginning of a long journey to insight, inner knowledge, and finally, to what Thomas Mann describes as "smiling knowledge of the eternal."[20]

After 20 years of wandering, led by his daughter Antigone, the old, blind, bearded, and ragged Oedipus seeks a final home in the sacred grove of the Furies at Colonus, the birthplace of Sophocles himself. Here, Oedipus realizes that he has come upon the place where the oracle predicted he would become a blessing to those who received him. Through his long exile, the elder Oedipus eventually sees that he has lived all of his life in accordance with the oracle's prophecy and that trying to evade this fate only intensified its power over him. His blessed death, Sophocles seems to be saying, is appropriate to one "who has lived long enough to understand the meaning of his own story."[21]

Written by an old man on the verge of Athens' defeat in its long war with Sparta, *Oedipus at Colonus* is the only Greek tragedy to center around an aged hero. Old people did not fare well in Greek tragedy, which emphasized the foibles of old age and portrayed aged men as either sinister or slightly ridiculous.[22] Certainly, the repulsive Oedipus—who first appears on stage as a blind, filthy, old beggar—seems unpromising as a tragic hero. Yet despite his physical frailty and poverty, Oedipus is far from weak and pathetic; he too is like a rock, as Sophocles describes him, "with storm waves beating against it from every quarter."[23]

Oedipus at Colonus centers on Oedipus' search for a final resting place, and on his transformation from an exiled, wandering beggar into a hero who confers both sacred and secular gifts to the community that receives him. Through this transformation, Sophocles boldly envisions the elder Oedipus as a tragic hero, while portraying aging as a moral and spiritual journey.[24]

When Oedipus and Antigone first appear on stage, unaware that they are trespassing in the sacred grove of the Furies, Oedipus' words reveal none of the overconfidence in his intelligence and ability that marked him as a younger hero.

Who will be kind to Oedipus this evening
And give alms to the wanderer?

He continues:

Suffering and time
Vast time, have been instructors in contentment.[25]

However, humility and acquiescence in human limitation are not the only qualities of the aged hero; they make the other virtues possible. Oedipus' eventual triumph will not take place without great effort and continued suffering; he must first win the acceptance of the Athenians.

When a stranger enters and orders them to leave the forbidden grove, Oedipus realizes that his journey is nearly over. He recognizes his final resting place (foretold by the oracle), prays to the goddesses, and declares:

I shall never leave this resting place.[26]

Consciously affirming his destiny, Oedipus has at last found the source of renewed strength, and his transformation begins. He asks to see Theseus, the king of Athens. For a small favor, he tells the skeptical stranger that Athens will receive a great benefit.

When the chorus of elders (guardians of the sacred grove) learn the wanderer's real identity, they revoke their promise to protect him and attempt to drive him out. They do not want their city polluted by his cursed presence. Oedipus replies with a powerful challenge to the Athenians' conventional piety; he takes the elders to task for failing to live up to the city's reputation as a refuge for distressed strangers.

For Athens, so they say, excels in piety;
Has the power to save the wretched of other lands;
Can give them refuge...
Yet when it comes to me, where is her refuge?
You pluck me out from these rocks and cast me out;
All for fear of a name.[27]

He insists that the events of his past were more suffered than performed; incomplete knowledge and lack of insight rather than moral depravity were the cause of his actions.

The integrity and power of Oedipus' challenge compels the chorus to reconsider. Whereas they believe themselves justified in banishing Oedipus because the gods have punished him, Oedipus warns them against presuming to know the judgments of the gods. True reverence for the gods, he implies, requires them to transcend their notion of him as polluted and to accept him as someone who has suffered deeply from the basic human flaw of imperfect knowledge. By comparing his reputation for evil to the Athenians' reputation for compassion, Oedipus strikes at the heart of Athens' moral ideals.

Oedipus' challenge so impresses the chorus that they call for the king. When Theseus enters, he immediately recognizes and welcomes Oedipus, affirming Athens' commitment to sheltering the stranger. Theseus too has lived as an exile; he grants Oedipus' wish to remain forever in Athens. Oedipus learns that his sons Eteocles and Polyneices are at war for the throne of Thebes. Since the oracle has decreed that victory will come to whomever controls Oedipus' dead body, Creon (who is allied with Eteocles) comes to Athens to try to take the old man.

Although Oedipus' moral integrity and inner strength win the Athenians' favor, neither they nor the chorus of elders can protect him against Creon's armed guards. The power of Theseus and his men is required to save Oedipus and his daughters from the designs of Creon and Eteocles. When Oedipus' other son, Polyneices, comes to plead for help, Oedipus reluctantly agrees to give him an audience, but mindful of his son's physical strength and his own frailty, Oedipus first asks Theseus to guarantee his safety.

Theseus departs, leaving two soldiers to protect Oedipus, who is sitting in the sacred grove with his daughters. The chorus sings a powerful, melancholy ode.[28] Written by an old man and sung by old men to an old man who is about to die, the poem focuses on the harsh realities of old age, framed by the conventional Greek understanding of the three ages of life.

Though he has watched a decent age pass by,
A man will sometimes still desire the world.
I swear I see no wisdom in that man.
The endless hours pile up a drift of pain
More unrelieved each day; and as for pleasure,
When he is sunken in excessive age,
You will not see his pleasure anywhere.

Old age, like youth and the prime of life, has its own natural boundaries and characteristics that must be accepted. Youth is a time of "feathery follies," whereas midlife carries with it the heavy burdens of

Jealousy, faction, quarreling and battle—
The bloodiness of war, the grief of war.

Finally,

...comes to strengthless age,
Abhorred by all men, without company,

Unfriended in that uttermost twilight
Where he must live with every bitter thing.

These sentiments cannot be attributed to the distance or vanity of youth. To Sophocles and the chorus, they are lived realities:

This is the truth, not for me only,
But for this blind and ruined man.

From this vantage point, death is a relief; excess of life brings only pain. The elders express a commonplace Greek aphorism that we suspect they only half believe: "not to be born" would be best of all, but once born, "second best is...to go back quickly whence we came."[29]

Oedipus' life might seem to validate such pessimism; he has suffered in all three ages of his life—as a child abandoned on a hillside, ankles pinioned together to prevent his crawling away; as a king reduced from greatness to misery; and as an old man wandering in exile. Yet the chorus expresses admiration as well as pity for his persistence and survival. In terms that recall the Sphinx's riddle, they compare him to a rocky, northern cape battered by storms:

so him too, ruinously
do terrible wave-breaking
calamities lash, never ceasing.

some from the setting of the sun
some from its rising
some through the midday beam
some from the night-wrapped north.[30]

Oedipus' sufferings continue until the end of his journey. He turns a deaf ear on Polyneices' plea and prophesies that the two brothers will die at each other's hands.

A loud clap of thunder from Zeus signals that Oedipus' time has come. Theseus, who now recognizes Oedipus' heroic, god-like status, arrives to receive instructions. To protect Athens from her enemies, he and his royal descendants alone may know the secret place of Oedipus' burial.

Mysteriously, Oedipus now leads the way to a place in Demeter's sacred grove where his daughters bathe and dress him in a ritual preparation for death. He declares his love for his daughters and bids them farewell. The chorus prays that his death will bring relief from suffering. A messenger declares that Oedipus

> *...was taken without lamentation,*
> *Illness or suffering; indeed his end*
> *Was wonderful if mortal's ever was.*[31]

Oedipus' long tragic road of life, then, culminates in his passage into an immortal god-like figure.[32] Although Oedipus literally becomes a "bridgehead to the sacred,"[33] Sophocles leaves no impression of a final triumph for Oedipus' family. In *Antigone,* Oedipus' daughters continue to live out the mystery of undeserved suffering and the inscrutable fate that marks the world of human temporality.

What, then, is the wisdom that enables Oedipus to transcend his tragic suffering and helplessness? As an aged hero, he has actually lived out the riddle of the Sphinx—the riddle of human existence in time. As a younger hero, Oedipus attempted to master time by uniting the three ages of human life in a single answer: man. His familial actions parallel this intellectual attempt to control time: patricide, which allowed him to take his father's place too soon, and incest, which effaced time by "collapsing the necessary temporal distinctions between generations."[34]

When Oedipus finally arrives at Athens and wins acceptance there, he has lived sequentially through all the terms of the riddle. Although his agonized self-blinding forces him to walk prematurely on "three feet,"[35] not until the end of his long journey of exile does he possess both the inner and the outer knowledge that mark him as an aged hero. When Oedipus has lived long enough to understand the meaning of his own story, he has acquired an existential knowledge that cannot be replaced by intellectual prowess. The riddle of the Sphinx cuts deeper than abstract truths about the human life cycle. "Not in pondering ideas but in surveying one's inner life and discovering the graveyard of needs and desires, once fervently cherished," writes Abraham Heschel, do "we become intimately aware of the temporality of existence."[36]

If "time, vast time, and suffering" have been Oedipus' "instructors in contentment," they have also shown him his insufficiency and his need for others. Oedipus needs both Antigone and Athens to become who he will be. He achieves this understanding by painfully fulfilling the injunction "Know Thyself," which is inscribed at the entrance of Apollo's temple at Delphi, but Sophocles takes us beyond the traditional Greek understanding of this maxim, i.e., know that you are human and mortal—not a god. The acquisition of self-knowledge, then, is central to the fulfillment of the aged hero's destiny. Personal knowledge, which includes knowledge of his own fatedness, allows the hero to come to a new appreciation of love and community.

In *Oedipus at Colonus,* we see the mutual redemption of the exiled aged hero and society. Reintegration into the community is a difficult task that demands delicate reciprocity. The aged hero must be capable of persisting on the tragic road to self-knowledge, whereas the community must be capable of tolerating the unknown, the fearfully alien. Oedipus' own personal integrity enables the Athenians to affirm their most cherished values. They affirm their openness and compassion by accepting him, not out of pity, but out of respect for an individual who has struggled with the conditions of his own existence.[37]

Sophocles, then, has portrayed aging as a moral and spiritual journey. Its surprises, terrors, mysteries, and triumphs, he implies, can only be successfully crossed with humility and self-knowledge, love and compassion, acceptance of mortality, and a sense of the sacred. Our own scientific culture resembles Oedipus as a young hero, the brilliant problem solver who neglects the existential ground of his greatness. Like the young Oedipus, we study the ages of life to control the facts of human existence in time and space, and like the young Oedipus, we have suppressed the existential terrain of the life cycle by construing it as a technical problem. As a result, old age in our culture too often seems like a "season" without a purpose; old people too often appear only as strangers, not as pilgrims.

Yet for all its power, the triumph of Oedipus remains troubling. Oedipus prevails, but in *Antigone,* his family line comes to a miserable end. Rather than forgive even one of his sons, he allows them to slaughter each other in battle. His daughters live out the mystery of undeserved fate. We are forced to wonder: did Oedipus' children pay the price for his glory? Did he put his own spiritual welfare above obligations to their future lives? Does successful culmination of life's journey require undue sacrifices from those, like Antigone, who make it possible?

The Greek tragedians insisted that all wisdom was incomplete, and every triumph exacted its toll. I do not know the answers to these questions. I do know that the aged Oedipus moves me to face my own destiny. Standing between my daughter and grandmother, I recognize myself as the animal that goes on four legs in the morning, two legs at midday, and three legs in the afternoon. Cut loose from the anchoring old rocks of my childhood, I embark on what Anne Sexton calls "the awful rowing toward God," in search of an innocent and playful wisdom. Along the way, I hope to see if my own old age corresponds to my middle-aged ideal of it.

Notes and References

[1]Since the mid 1970s, a new literature has grown up at the intersection of gerontology and the humanities. *See* D. Polisar, L. Wygant, T. R. Cole, and C. Perdomo (1988) *Where Do We Come from? What Are We? Where Are We Going? An Annotated Bibliography of Aging and the Humanities.* Gerontological Society of America, Washington, DC; also, T. R. Cole, D. D. Van Tassel, and R. Kastenbaum, eds. (1991) *Handbook of Aging and the Humanities.* Springer Publishing Company, New York, NY.

[2]*See* T. R. Cole (1984) The prophecy of *Senescence:* G. Stanley Hall and the reconstruction of old age in America. *The Gerontologist* **24 (4),** 360–366.

[3]L. Kass (1983) The case for mortality. *American Scholar* **52 (2),** 173–191.

[4]G. S. Hall (1922) *Senescence.* D. Appleton, New York, NY, p. 403.

[5]C. G. Jung (1933) The stages of life, in *Modern Man in Search of a Soul.* Harcourt Brace Jovanovich, New York, NY, pp. 95–114..

[6]A. L. Vischer (1967) *On Growing Old* (G. Onn, trans.) Houghton Mifflin, Boston, MA, p. 23.

[7]E. Erikson (1964) Human strength and the cycle of generations, in *Insight and Responsibility.* Norton, New York, NY, p. 132.

[8]I. Rosow (1974) *Socialization to Old Age.* University of California Press, Berkeley, CA, p. 148.

[9]L. Rosenmayr (1980) Achievements, doubts, and prospects of the sociology of aging. *Human Development* **23,** 60.

[10]H. R. Moody (1986) The meaning of old age and the meaning of life, in *What Does It Mean To Grow Old? Reflections from the Humanities* (T. R. Cole and S. Gadow, eds.), Duke University Press, Durham, NC, pp. 11–17.

[11]M. N. Fried and M. H. Fried (1980) *Transitions, Four Rituals in Eight Cultures,* Norton, New York, NY; J. Keith and D. Kertzer, eds. (1984) *Age in Anthropological Theory,* Cornell University Press, Ithaca, NY; M. Furlong (1973) *The End of Our Exploring,* Hodder and Stoughton, London, UK; G. Roppen and R. Sommer (1964) *Strangers and Pilgrims,* Norwegian Universities Press, Bergen, Norway; J. S. Dunne (1973) *Time and Myth,* Doubleday and Co., New York, NY.

[12]*See* L. Nash (1978) Concepts of existence: Greek origins of generational thought, *Daedalus* **107 (4),** 1–22. *See also* the fable of Aesop, which combines primitive Germanic and Oriental ideas about the life cycle:

> God wanted man and the animals to have the same length of life—thirty years. But the animals observed that this seemed too long a time to them, while it seemed very little to man. Then they came to an accord, and the donkey, the dog, and the monkey agreed to hand over part of their years to those of man. In this way the human being achieves a life of seventy years. The first thirty are good years, and in them man enjoys good health, he amuses himself, he works joyfully, happy in his destiny. But then come the eighteen years of the donkey, when he has to bear load after load, he must carry the grain which someone else eats, and endure kicks and beatings in return for his good service. Then come the twelve years of the dog's life; he crawls into a corner, growls and shows his teeth, though he has very few teeth left for biting. And when this has passed, then come the ten years of the monkey, which are the last; man makes whistling noises and foolish gestures, busies himself with absurd manias, goes bald, and is useful only as a butt for the laughter of little children.

Cited in Ortega y Gasset (1958) Idea of the generation, in *Man in Crisis,* Norton, New York, NY, p. 48.

[13]Cited in J. A. Burrow (1986) *The Ages of Life,* Oxford University Press, Oxford, UK, Appendix, pp. 191–194. The three ages of life was not the only age-grading scheme in ancient Greek culture. Other divisions mingled easily with one another, each pertaining to a particular context—e.g., heroic, agricultural, domestic, erotic, biological, or medical. The physical, behavioral, ethical, and social characteristics ascribed to each "age" depended on the scheme's context. Homeric epic, for example, depicted four ages of life (childhood, young manhood, mature manhood, old age), each shaped according to the demands of heroic warfare. Solon's numerically defined ten ages, though unfamiliar in Greek literature, reflect commercial concerns in sixth century Athens and foreshadow the rationalization of the life course in modern society. *See* T. Falkner (1988) The politics and the poetics of time in Solon's "Ten Ages." Unpublished paper. I am grateful to Professor Falkner for information about the linguistic and cultural conventions of the life cycle in ancient Greece, personal communication, June 8, 1988.

[14]Burrow, *The Ages of Life,* p. 1

[15]A. Heschel (1972) *Who Is Man?* Stanford University Press, Stanford, CA, p. 39.

[16]Furlong, *The End of Our Exploring,* pp. 19,20. Traditionally of course, the journey motif applied only to masculine heroes—and younger ones at that. In C. Ochs (1983) *Women and Spirituality* (Rowman and Allanheid, Totowa, NJ), Ochs argues that the journey metaphor in Western spiritual writings contains a male bias that does not accord with women's experience. Although this is an important criticism, I believe she has put the case too strongly—missing, for example, the journey of Christiana in Bunyan's (1684) *Pilgrim's Progress,* Part Two. Recent feminist writers and theologians, on the other hand, have used the motif for their own purposes. *See,* for example, N. Morton (1985) *The Journey Is Home,* Beacon Press, Boston, MA; and C. P. Christ (1980) *Diving Deep and Surfacing: Women Writers on the Spiritual Quest,* Beacon Press, Boston, MA. For recent usage of the journey metaphor in religious writings on later life, *see* G. O'Collins (1978) *The Second Journey,* Paulist Press, New York, NY; and E. Bianchi (1984) *Aging as a Spiritual Journey,* Crossroads, New York, NY.

[17]Roppen and Sommer, *Strangers and Pilgrims,* pp. 18–20.

[18]Heschel, *Who Is Man?*

[19]N. Datan (1985) The Oedipus cycle: Development mythology, Greek tragedy, and the sociology of knowledge. Paper presented to the Gerontological Society's annual meeting, New Orleans, LA; J. M. Ross (1988) Oedipus revisited: Laius and the "Laius Complex," in *The Oedipus Papers* (G. H. Pollock and J. M. Ross, eds.), International Universities Press, Madison, WI, pp. 235–316.

[20]Cited in C. Downing (1981) Your old men shall dream dreams, in *Wisdom and Age* (J. R. Staude, ed.), Ross Books, Berkeley, CA, p. 172.

[21]Downing, *Wisdom and Age,* p. 176.

[22]B. M. W. Knox (1964) *The Heroic Temper,* University of California Press, Berkeley, CA, p. 145; B. E. Richardson (1969) *Old Age Among the Ancient Greeks,* AMS Press, New York, NY (originally published in 1933).

[23]Knox's translation, *The Heroic Temper,* p. 9.

[24]A. C. Schlesinger (1953) Tragedy and the moral frontier. *Transactions and Proceedings of the American Philological Society* **84,** 164–175.

[25]Sophocles (1977) Oedipus at Colonus in *Sophocles: The Oedipus Cycle* (D. Fitts and R. Fitzgerald, eds.), Harcourt, Brace, and Jovanovich, New York, NY, p. 82.

[26]Sophocles, *Oedipus at Colonus,* p. 84.

[27]Sophocles, *Oedipus at Colonus,* p. 95.

[28]*See* G. H. Gellie (1972) *Sophocles,* Melbourne University Press, Melbourne, Australia, pp. 273–279, for a reading of this poem.

[29]On Greek images and ideas about death, dying, and the dead, *see* R. Garland (1985) *The Greek Way of Death,* Cornell University Press, Ithaca, NY; E. Vermeule (1979) *Aspects of Death in Early Greek Art and Poetry,* University of California Press, Berkeley, CA.

[30]Gellie, *Sophocles,* p. 274.

[31]Sophocles, *Oedipus at Colonus,* p. 163.

[32]On the cult status of Greek heroes, *see* R. Scodel (1984) *Sophocles.* Twayne, Boston, MA, pp. 111-112; C. Segal (1981) *Tragedy and Civilization: An Interpretation of Sophocles.* Harvard University Press, Cambridge, MA, p. 364 in Euben.

[33]On the role of old men as "bridgeheads to the sacred" in traditional cultures, *see* D. Gutmann (1981) Observations on culture and mental health in later life, in *Handbook of Mental Health and Aging* (J. Birren and R. Sloane, eds.) Prentice-Hall, Englewood Cliffs, NJ, pp. 429–447. In this light, the chorus of elders' role in guarding the sacred grove of the Furies takes on new significance. *See* R. W. B. Burton (1980) *The Chorus in Sophocles' Tragedies,* Clarendon Press, Oxford, UK. For a recent interpretation of *Oedipus at Colonus* along these lines, *see* T. V. Nortwick (1989) "Do not go gently..." *Oedipus at Colonus* and the psychology of aging, in *Old Age in Greek and Latin Literature* (T. M. Falkner and J. de Luce, eds.) State University of New York Press, Albany, NY, pp. 132–156.

[34]Froma I. Zeitlin (1986) Thebes: Theatre of self and society in Athenian drama," in *Greek Tragedy and Political Theory* (J. P. Euben, ed.), University of California Press, Berkeley, CA, p. 128.

[35]Zeitlin, in Euben, *Greek Tragedy and Political Theory,* p. 129.

[36]Heschel, *Who Is Man?* p. 62.

[37]L. Slatkin, Oedipus at Colonus: exile and integration, in Euben, *Greek Tragedy and Political Theory,* p. 217.

Recovering the Body in Aging

Sally Gadow

The relation between self and body is pivotal in aging. The physical alteration that accompanies growing old is commonly regarded as one of its most problematic aspects. The loss that is threatened with aging is not the sheer fact of physical decline, but the inevitable alienation from a body that is regarded as an imprisoning object. Such a body is uninhabitable, already a crypt from which death is the only release.

The extreme expression of that alienation is the hopeless contradiction deBeauvoir describes between "the inward feeling that guarantees our unchanging quality and the objective certainty of our transformation."[1] In the imagery of Bradbury's 90-year-old Miss Loomis, "a body like this is a dragon, all scales and folds. So the dragon ate the white swan. I haven't seen her for years...I *feel* her, though. She's safe inside, still alive."[2] With a final, desperate image, she adds, "I'm the princess in the crumbled tower, no way out." Blythe's elderly subject expresses the same anguish: "The old part of me worries the young part of me. It could be that it would be better to be all old."[3]

The origins of the self–body antagonism that grows so poignant in aging are to be found not in the physical changes of aging, but in the earliest consciousness of the body as an object of meaning. In order to envision other relations between self and body beyond the impasse expressed by deBeauvoir, Bradbury, and Blythe, it is necessary to examine those origins.

Aging and Ethics Ed.: N. Jecker

The body is the object of distinct meanings that influence an individual's experience at any age. The older person's experience represents a continuation—often a tragic culmination—of that influence rather than a peculiarity of advanced age. Significantly, contemporary meanings that inform the self–body relation are not a coherent system, but a conjunction of antithetical values. The body is esteemed as the most important, even the most real, aspect of a person's being, while it is also disdained as the corrupt, nonessential element of existence—a facade for the true essence of the person.

The destructiveness of these views is displayed compellingly in Albee's drama, "The American Dream," in which the body of the Young Man is his social currency, his sole validity. Grandma offers him the ultimate accolade: "Boy, you know what you are, don't you? You're the American Dream."[4] Applying the standard of the dream-body to Grandma's physical qualities, the result might be termed the American nightmare: "I look just as much like an old man as I do like an old woman."[5] However, the Young Man's body too is a horror; he is, metaphorically, a castrate. "[A]fter they cut off its you-know-what, it *still* put its hand under the covers, *looking* for its you-know-what. So, finally, they *had* to cut off its hands at the wrists."[6] The value of his body has been annulled by the judgment of it as corrupt, yet in the opposing ideology, it remains the only part of his being that is valued or real—as real, at least, as a dream.

The Body as Object

The body of Albee's Young Man is a dream in both senses of the word. It is the instance of an ideal, and it is fictive, an image dreamed for him by others. It is, in short, a social construct rather than a part of the self.

With the earliest consciousness of the body as object of social meanings, the ground is prepared for irreconcilable antagonism between self and body in aging. Although the antagonism may be more characteristic of aging than of youth, the alienation from the body that normative meanings produce is present whether the body be a dream or a nightmare. The body of the Young Man conforms utterly to social requirements, yet it is as remote from the self as is the dragon from the white swan of Miss Loomis. For both persons, the meaning and value of the body are externally decided; they are not given originally by the self.

In addition to the mediation of external meanings, the body acquires an immediate objectness in the encounter with physical limitation. Pain, illness, and fatigue are intrinsic to physical existence. As long as the body does not obstruct the projects of the individual, its meaning as a physical object is insignificant, but when the objectness takes concrete form as an impediment, it is no longer background; it becomes central and so demands a meaning.

One meaning the body can have is pure object, losing all lived connection with the self.

> I watched the nurse move it, and it was like watching someone else's leg, or a piece of furniture, being moved. My own leg, in this moment, was no longer my own, and no longer part of my body or my self. No longer part of *anything*...Suddenly, then my leg (my ex-leg) was Nothing and Nowhere, so far as living or felt reality was concerned; ...it had gone, and left not a trace behind.[7]

Alternately, the body whose frailty dominates experience can be viewed not as a neutral Nothing, but as a betrayer, an enemy with whom one must continually make peace.

> Your body is suddenly a being apart from you, a traitor...you have to get its permission to take it on walks...you have to give it naps, feed it proteins, promise it that it will get over its hurt feelings.[8]

On the other hand, constraints of the body as physical object may not be experienced as alienating. "Perfect brightness...even exuberance of the spirit, is compatible in my case," insisted Nietzsche, "not only with the most profound physiological weakness, but even with an excess of pain."[9] In his last writing, only weeks before his collapse, he still affirmed the "sweetening and spiritualization which is almost inseparably connected with an extreme poverty of blood and muscle."[10]

The Person as Object

Meanings of the body as physical and social object converge when the body as object becomes the person as object. The experience of medical patients elucidates this, providing a key to the more subtle cultural reduction of older persons to the status of objects.

> I had a sense of the comically close parallel between the way I regarded the leg, and the way the surgeon regarded me. I regarded the leg as "a thing," and he, apparently, regarded me as "a thing." Thus I was doubly "thinged," reduced to a thing with a thing.[11]

The old, like the disabled, are doubly thinged, but in a more comprehensive manner. They are objects of the same ambivalent treatment that is given any worrisome minority—economic discrimination, social beneficence, and above all, scientific scrutiny; they finally may be rendered so special as to be considered a separate species. However, a more radical reduction occurs in the very categorization of persons as old. Being old, like being alluring or loathsome, expresses an external judgment that the self takes up as its identity but does not first experience as an inward reality with the immediacy of pain, energy, or joy. "Old" is the meaning used to organize and interpret various phenomena; it is not the description of a self-evident state that individuals find in themselves. Indeed, the realization that one is regarded as old can seem a shocking absurdity bearing no relation to the person's sense of self. "It surprises me to be so old," confesses one of Blythe's subjects.[12] Why? The category of old age is, in Sartre's term, "an unrealizable." It is impossible to experience, from within, one's identity for others. This applies not only to aging, but to any characterization of the individual by others. The body as social object is one example. "The unrealizable is my being seen from without...something of which I cannot have any full inward experience."[13]

The effect on the individual of social images of the body can be kept within bounds, since the body is only part of one's reality (the white swan is untouched by the dragon). However, an attribute like "old" is intended as an encompassing designation of the person's entire being. An ascription that is meant to refer to the whole can be transcended only by the most severe measures: a young or timeless being is posited beside the old one. So divisive is the attempt to preserve subjective reality against the pressure of an inclusive category like old that, indeed, it might be better for some to be "all old." The inexorable logic of total categories, like the power of total institutions, obliterates the individual's freedom of self-definition, in that way negating the fundamental difference between persons and objects. According to that logic, it is impossible to be both old and not-old, to be partially old. The logic applies to the not-old as well. Albee's character is distilled without remainder into "Young Man," nameless. He is reduced to a category and, thus, to an object, so far from being an individual that "Young" not only describes him but names him.

The reduction of persons to objects is a profound moral problem for ethical positions in which respect for persons is a value. However, it is a serious existential problem as well, in terms of the destructive outcome when individuals regard themselves as objects.

That destructiveness can take two forms. First, the alienation of an individual from the body imperils the person's connection with the world, a connection the body provides. An example is the nursing home resident who refuses to eat in the company of others because she fears their disgust at her disability. Physical disengagement also may be one corollary of an expectation that older persons be, as it were, disembodied, bearers of spiritual wisdom who have transcended physical exigencies. In addition to social isolation, a second consequence of alienation from the body is disintegration of the individual: persons lose the connection not only with their world, but with themselves. Even in its most revolting, recalcitrant state, the body is still an anchor for the self, its most palpable reality. Disembodied—without a lived relation to the body—we can only worry, with the poet, "how we shall call/Our natures forth when that live tongue is all/Dispelled, that glass obscured or broken/... the singing locust of the soul unshelled."[14]

Recovery of the Body

Loss of the body occurs through its reduction to a social image or to a physical other, or both—in short, reduction of part of the person to the status of object. Reduction of the entire person to an object follows easily in a culture characterized by an obsession with the body, particularly for the person whose body is a supreme fulfillment or a radical violation of the standards for bodies. One means of overcoming the dangers, moral as well as existential, of reducing persons to objects is to recover the body as part of the self, to reconcile its objectness with the subjectivity of the person.

One form of reconciliation between self and body that conventionally has been denied the elderly is sexuality. The body that becomes a dragon is a body that no one (except another dragon) would desire. Whereas the place of sexual fulfillment in the hierarchy of human values can be debated, the serious fragmentation of the person that results when sexuality is forbidden cannot be disputed. For the older person whose relation to the body may already be in question, the result can be brutal.

> Modern geriatric psychiatry speaks of the aged as being wounded in their narcissism—a poignant term and one which eloquently compresses the whole business of what we once were and what we must inevitably become, should we see our time out. To be sexually prohibited, even if it is only in some ignorant, unwritten and folk-saw sense, is a gratuitous addition to this hurt.[15]

Another means of recovering the body is an emphasis on its subjective uniqueness. The estrangement of the body as a social object occurs through the mechanism of standardization. Categories and criteria—cosmetic, athletic, or clinical—are applied to every feature of the body, inner and outer, dictating the optimal size, shape, serum cholesterol level, and the like, that bodies are to have. Eventually, all physical phenomena come to be interpreted via external standards, to the point that experience is acknowledged only when it can be comprehended through one of the accepted categories. The suffering, for example, that accompanies an inability to control excretion or to move freely is overlooked, even disallowed, by the diagnosis "aging." Scott-Maxwell grieves, "Real pain is there, and if we have to be falsely cheerful, it is part of our isolation."[16] The isolation is twofold: one becomes estranged both from others and from the body whose subjective reality others deny. The view that emphasizes uniqueness and subjectivity over external standards is captured exactly by the protest, "There is more to us than age."[17] The "more" is not limited to suffering. Blythe reminds us that

> ...the old often have amused eyes and are not necessarily desperate. Serviced with dentures, lenses, tiny loudspeakers, sticks and hip-pins, the flesh has become absurd and can no longer be taken seriously. The body has become a boneshaker which might just about get you there if you are lucky.[18]

The boneshaker is absurd, but not an enemy of the self. On the contrary, it is the opportunity to cultivate a deliberate integrity of self and body, to cherish and not renounce the body, to care for it as one would a beloved with whom one has laughed and danced and from whom one is soon to be parted.[19] Frailty becomes dialectical, at once a limit and a freedom—the freedom to create a self–body relation in which the body is not a mere object, but a subject, an old/new companion of greatest intimacy.

References

[1]S. deBeauvoir (1972) *The Coming of Age* (P. O'Brian, trans.), G. P. Putnam's Sons, New York, NY, p. 290.

[2]R. Bradbury (1964) *Dandelion Wine*. Bantam, New York, NY, p. 109.

[3]R. Blythe (1979) *The View in Winter: Reflections on Old Age*. Penguin, New York, NY, p. 188.

[4]E. Albee (1961) *The American Dream.* New American Library, New York, NY, p. 108.

[5]Albee, p. 111.

[6]Albee, p. 100.

[7]O. Sacks (1982) The leg. *London Review of Books* **17,** 3.

[8]V. Weingarten (1978) *Intimations of Mortality.* Alfred A. Knopf, New York, NY, p. 89.

[9]F. Nietzsche (1968) Ecce homo, in *Basic Writings of Nietzsche* (W. Kaufman, ed. and trans.) Modern Library, New York, NY, p. 78.

[10]Nietzsche, p. 78.

[11]Sacks, p. 4.

[12]Blythe, p. 169.

[13]deBeauvoir, p. 291.

[14]R. Wilbur (1963) Advice to a prophet, in *The Poems of Richard Wilbur.* Harcourt, Brace, Jovanovich, New York, NY, p. 7.

[15]Blythe, p. 9

[16]F. Scott-Maxwell (1979) *The Measure of My Days.* Penguin, New York, NY, p. 32.

[17]Scott-Maxwell, p. 140.

[18]Blythe, p. 212.

[19]S. Gadow (1983) Frailty and strength: The dialectic in aging. *The Gerontologist* **23,** 144–147.

Aging and Filial Responsibility

The Aging Society as a Context for Family Life

Gunhild O. Hagestad

The focus of this chapter is on how families and relationships within them have been affected by demographic change and by the wider societal changes that are associated with population aging. Decreased mortality, altered patterns of fertility, and the increasing imbalance in sex ratios during the second half of adulthood have had profound effects on families. This chapter relates these changes to some key aspects of family life, touches on how other characteristics of aging societies have contributed to changes in family worlds, and outlines some of the issues that will be confronting families as our aging society faces the twenty-first century. A final section contrasts the consequences these patterns of change are posing for the separate life experiences of men and women.

We shall be looking especially at three main trends in recent population change. The first is the increased general life expectancy and what demographers refer to as "the rectangularization of survival curves"—that is, the fact that the majority of the population now survives into old age, and about three-fourths of all deaths occur among those over the age of 65. The second trend is the decrease in family size and the smaller age difference that prevails today between a typical family's oldest and youngest child. Around the turn of the century, mothers bore, on the average, 3.9 children. The current estimate is about 1.8. The widened gap between the mortality rates of men and women, which has produced a seven-

Aging and Ethics Ed.: N. Jecker

year difference in their general life expectancy, is the third major change. Together, they pose some often surprising challenges and possibilities for family life in the near future.

Demographic Change and Family Life

Altered Patterns of Family Deaths

> ...We place dying in what we take to be its logical position, which is at the close of a long life, whereas our ancestors accepted the futility of placing it in any position at all. In the midst of life we are in death, they said, and they meant it. To them it was a fact; to us it is a metaphor.[1]

British author Ronald Blythe has captured the human realities behind the demographers' life tables. Less than a century ago, death was a normal experience in all phases of family life; only in the last few decades has it become associated more and more with advanced age.

Recent work on turning points in the course of human lives has emphasized the need for predictability—the craving to know what lies ahead.[2] Those life changes that we regard as normal, and thus expect, do not typically become crises, but unanticipated or untimely events sometimes do. When events come at a "scheduled" time, they are less likely to catch us unprepared than when they are unscheduled or come too early. Life changes are also easier when they occur "on schedule" because the individuals who are experiencing them may enjoy the comfort of others who are "in the same boat."

Recent changes in mortality have made the time of death more predictable and have clarified the meaning of the term "untimely death." Today, the death of parents before their child has reached midlife, and the death of their children—at any age—would be assigned that label. Because such events are not expected, they may be more traumatic now than they were in the past, for they catch us in the vulnerability of unpreparedness. They are also likely to be lonely transitions, neither shared nor fully understood by peers.

Among children born in 1910 who survived to their 15th birthday, more than one-half had experienced the death of a parent or a sibling by their early teens.[3] Today, the loss of parents is not expected until the second half of adulthood. For current generations of young women, the death of the mother may come close to the daughter's retirement age.[4]

At the beginning of this century, parents with an average number of children had a .62 chance of experiencing the death of a young child. In the late 1970s, the probability was only .04. The loss of a child has ceased to be a normal, anticipated part of family life.

For a growing number of adults, it is the loss of a grandparent that constitutes their first encounter with family deaths. We now anticipate that the death of grandparents will not occur until our early adulthood, but until quite recently, most grandparents did not live long enough to have significant life overlaps with their grandchildren.[5]

Concern is growing over the strain experienced by families who provide care for their ill or impaired elderly members. It is frequently argued that families in an aging society must be prepared to assume more of a care load than was the case for families in the past. We do not really have the historical data to judge if such statements are accurate, but it is important to keep in mind that, before improved living conditions and medical advances produced rectangular survival curves, illness and death were encountered in all phases of family life—and at a time when there were far fewer outside institutional facilities and programs to alleviate their pressures. Families have always been caregiving units that were expected to experience and absorb the shocks of illness and bereavement. The main difference between today's families and families of the past is not likely to be in the total *amount* of care and concern they expend, but the *focus* of them.

Infants and young children have always been regarded as vulnerable and dependent. What is new about the current demographic picture is that, after childhood, these attributes are linked nearly exclusively with old age. It is safe to say that never before in human history has the experience of human frailty and the loss of a family member been so clearly linked to one group: the old. Perhaps part of the sense of burden associated with their care stems from the fact that the illness and loss are represented by *parents*—individuals who for decades were perceived as pillars of strength and support.

Durable Bonds

Today's increased longevity provides new opportunities for individuals to accumulate a variety of experiences and skills. The same theme emerges when we consider decades of joint survival in family relationships across generations. Parents and children may accrue half a century of interwoven biographies, shared experiences, and memories. For the majority of life years shared by parents and children, the child is not *chronologically* a child. As we were growing up, most of us heard our parents say: "Some day you will understand," or "When you are a parent yourself...." Research on adults, particularly on mothers and daughters, bears out the folk wisdom: there seems to be new rapport and empathy between parents and children once the children become parents themselves.

In an aging society, a growing proportion of parents and children will share such key experiences of adulthood as work, parenthood, and even retirement and widowhood. In the grandparent–grandchild relationship, the grandparent may survive to experience many years of the grandchild's adulthood.

There is little doubt that altered mortality and fertility patterns have created a new climate for the building and maintenance of family relationships. Historians and demographers have suggested that, under conditions of high mortality, people were reluctant to form strong attachments to particular individuals, knowing that such ties could not be counted on to endure. Instead, they invested in the security of the family as a group. Today, most people not only take long-term family bonds for granted and invest accordingly in them, but, because of reduced fertility rates, there are also fewer individuals within each generation to invest in. As a result, intergenerational relationships are not only more *extensive* now, but they may also have become more *intensive.* Such "intensification of family life" may have gained even further momentum because of the weakening ties between family and community.[6]

A New Uniformity in Childhood Experiences?

We have noted that, in recent decades, average family size has declined, as has the age difference between the oldest child and youngest child in the family. In some of my own work, I have used Norwegian census material to examine the types of families that firstborns grew up in during various decades of this century. Among Norwegian children born during the 1920s, nearly one-half grew up in families of five or more children, and close to one-third were 16 or older when their youngest sibling was born. Among firstborns from the 1950s, only one in seven ended up with four or more siblings, and less than one-tenth of this group have 16 years or more between them and their youngest sibling. The proportion of firstborns who grew up with one or two siblings went from one-third in the 1920s group to nearly two-thirds in the 1950s group. These changes suggest an increasing uniformity in childhood experiences and in parent–child relationships as well.

Over the last century, successive generations of children have become more similar in terms of birth order, and the very meaning of sibling position may have changed. Among children born in 1870, one-third were either first or second children in their families. For children born in 1930, this proportion had nearly doubled.[7]

Psychology offers a sizeable amount of literature on the effects of sibling position on individual development. Many of these generalizations regarding birth order may not hold true when siblings are less than five years apart and come to think of themselves as peers once they reach adulthood. This is in sharp contrast to siblings from families in which the mother bore children throughout her fertile years, and firstborns were adults and parents themselves before the youngest sibling was born. In such cases, common until very recently, oldest and youngest siblings grew up in different families and encountered their parents in strikingly different life phases.

Furthermore, the experience of parental death was also likely to occur in quite different phases of life for the first- and lastborn, and they grew up with different "kin galleries." For one thing, firstborns had a much greater chance to know their grandparents than did their younger siblings.

As a growing proportion of children grow up in small families, other differences in their early life experiences are becoming reduced. There is strong evidence to suggest that small and large families constitute qualitatively different developmental contexts. Children in small families receive more attention from parents than do children with a number of siblings, and recent research has linked such differences to intellectual development and school performance.[8]

The Changing Age Structure of Kin Networks

Not only do the size and age-structure of nuclear families change in the process of population aging, but kin networks also take new forms. Four such changes are worth noting: multi-generational families are becoming more common; families are increasingly "top heavy;" family members have a growing number of vertical, intergenerational relationships, whereas relationships *within* generations are decreasing in number; and generational demarcations are becoming clearer.

Currently, about one-half of all individuals 65 or older are great-grandparents—that is, members of four-generation families. We have no good estimates of how common five-generation families are, but they, too, have most likely increased over the last few decades. The number of generations in a family lineage gives us an indication of the web of relationships it contains. In a four-generation family, there are three sets of parent–child connections, two sets of grandparent–grandchild ties, and members of two generations who occupy the roles of both parents and children. Individuals in a multi-generational family line relate in a much more complex set of family identities than is the case in a lineage with only two generations, those of parent and child. Under earlier demographic conditions, individuals were likely to have a variety of intragenerational, that is, horizontal, family ties. Under current conditions, a variety of relationships are *vertical,* crossing generational lines. An extreme illustration of such change would be China's recent one-child policy. If it succeeds and continues, Chinese in the twenty-first century will have only vertical family relationships outside the marital role, since they will have no siblings and no cousins.

Because of changes in fertility and mortality, the relative balance between old and young has shifted—in society as well as family. The demographer Samuel Preston reminds us that this is the first time in history that the average married couple has more parents than children.[9] There is little doubt that the age composition of family reunions is strikingly different now from the way it was at around the turn of the century. Then, it was not unusual to find families in which children and grandchildren were the same age, and aunts and uncles were younger than their nieces and nephews. With the period of childbearing now concentrated in a shorter span of women's adult years, and the age difference between first- and lastborn children reduced, generational demarcations have become clearer. Active involvement in the day-to-day demands of child rearing is also now likely to be over by the time women become grandmothers. The extent to which this has made grandmotherhood a more distinct role has not been fully explored, but it may mean that grandparents will increasingly emerge as a major stabilizing force in families, a point to which we shall return.

Both these trends—distinct generational demarcations and the clearer sequencing of parenthood and grandparenthood—are more typical of women's family lines than they are of paternal lines. Because of trends in age at first marriage and remarriage, and the growing number of "second families," men are more likely than women to have families whose age composition is characteristic of families early in this century.

Effects of "The Mortality Gap"

Women in the United States currently outlive men by about seven years. The world of the very old is a world of women, both in society and within families. Men and women also spend the latter part of their lives in differing living arrangements and relationships. Most older women are widows living alone; most older men live with their wives. For example, among individuals over the age of 75, two-thirds of the men are living with a spouse, whereas less than one-fifth of the women are.

These contrasts between older men and women have strong implications for the rest of their family members. First of all, the oldest members of a family are likely to be women. Women are also more likely to have great- and great-great-grandchildren. In societies where historical events have made sex ratios even more imbalanced than in this country, the three oldest generations may be populated only by women. For example, a German study found that many five-generation families contained three generations of widows.[10]

Differences in widowhood and remarriage mean that men tend to maintain a significant horizontal, intragenerational relationship until the end of their lives; women do not. Consequently, women draw more on their intergenerational relationships for help and support in old age. At the time when men face serious impairment, they are likely to have a wife to care for them, whereas frail and ill older women are typically widows who turn to younger generations for help. This may be one reason why women, throughout their adulthood, invest more time and energy in intergenerational ties than men do.

Of course, the lives of men and women have been affected by complex social and cultural processes that range far beyond demographic change. The extent to which this intensified the differences between the genders will be explored in a later section, but let us first consider some social changes that are associated with aging societies, and how they have affected family life.

The Impact of Wider Societal Change

The Lack of Cultural Rules

Demographic change may have been so rapid and so dramatic that we have experienced "cultural lags." Some of us find ourselves in life stages for which our society has no clear culturally shared expectations, and family members often face each other in relationships for which there is no historical precedence, and therefore, minimal cultural guidance on which to rely. Such themes are very clear in recent discussions of grandparenthood. Studies of grandparenting styles find a great deal of variety, and when a student and I analyzed popular magazine descriptions of grandparents in the 1880s and in the 1970s, we found current magazines to reflect a new uncertainty. Members of several generations expressed confusion over what grandparents were supposed to do. What are their rights and obligations? In a society where grandparents range from 25-year-olds to centenarians, and where grandchildren run the spectrum from infants to retirees, we should not be surprised to find that cultural images of grandparenthood harbor both variety and uncertainty.[11]

New Forms of Interdependence

Closely related to the issue of what people expect of one another in family relationships is the question of what holds them together. Over the last century, economic needs have given way to emotional needs as the main family "glue," especially in relationships among adults. Over the same time period, we have seen a shift in emphasis from the needs of the family as a group to the needs and wishes of individuals. Individual choices, such as decisions about when to marry and leave the family, were once guided by the needs of the family unit as a whole. The twentieth century, with its pension and health-care plans, has "freed" generations from many of these economic interdependencies.[12] It is commonly argued that, as a result, family ties have become more *voluntary* in nature.[13] Kin connections are seen as a latent potential, from which active and viable relationships may or may not develop.

Even though we have a good deal of cultural ambiguity regarding relationships between young and old in the family, members of our society still share some key values and norms about family responsibilities and interconnections. A moral imperative may be followed regardless of personal affection, which may explain why research has found no consistent relationship between rates of contact and feelings of emotional closeness between adults and their parents.[14] Surveys that have compared the attitudes and expectations of parents and children have often found that children are more ready than parents to state that the younger generation should provide help to needy elderly parents. It has also been found that the old are the most receptive to formal, nonfamily services, whereas the young are those most in favor of family provided help. Researchers attribute such contrasts to "youthful idealism" on the part of the young. The middle-aged and the old, on the other hand, are often responding on the basis of actual caretaking experiences. Recently, a number of writers have argued that, with the growth of societal supports for the old, an increasingly important function of the family will be to serve as mediators between bureaucracies and the aged,[15] and that modern families not only meet needs, but they identify needs, so that other institutions can address them.

Although pension systems and health plans have lightened the economic pressures for most families, there is still a steady flow of intergenerational support, and the majority of the states have enacted so-called "family responsibility" laws, statutes that establish relatives' responsibility for family members who are indigent, needy, or dependent. However, recent research and public debate indicate that enormous complexity still remains in sorting out rights and obligations among family members in an aging society.

Emerging Issues

Who Has the Right to Be Old?

The recent German study of five-generation families revealed that high levels of role ambiguity and strain confront the contemporary multigenerational family. Nearly 90% of the oldest generation were living in private households, and among these great-great-grandparents, 50% lived with a daughter who herself was in advanced old age. The families expressed a good deal of confusion about who had the right to some of the privileges of old age, and whose duty it was to provide them. The interviews often reflected a strong sense of jealousy in the two generations below the oldest, where members felt that they were missing out on some returns from their earlier family investments. The investigators warn that we may see increased rates of illness among young-old women in such families, both because of the stress of caring for the very old, and because of the possible benefits derived from the sick role, since the anticipated rewards of old age are not being realized. Perhaps more than any other society, Japan illustrates the potential ambiguity and strain produced by demographic change. Traditionally, the oldest son and his parents have maintained a pattern of coresidence within the Japanese stem family. With dramatic increases in life expectancy, the average duration of such coresidence nearly doubled between 1930 and 1950.[16] In today's Japan, an increasing number of families find themselves with two gray-haired and retired generations under one roof, both of whom expect support and deference.

In the United States, as in most Western industrialized societies, a good deal of debate has centered on families as units of caregiving. A focal point in such discussions has been the role of women in family networks and in intergenerational patterns of support.

Kin-Keeping and Its Costs

There is a rather extensive amount of literature showing that women are kin-keepers, and that their preparation for this role starts early in life. Kin-keeping tasks include maintaining communication, facilitating contact and the exchange of goods and services, and monitoring family relationships. These functions are often performed for the husband's kin as well as for the women's own family line. Even when they are not the initiators and orchestrators of family get-togethers, old women may nevertheless facilitate family contact by serving as the "excuse" for bringing kin together. The mother–daughter connection has emerged as the pivotal link, both in the maintenance of family contact and in the flow of support.

Daughters have been found to be the linchpin of widows' support systems. When aging parents live with offspring, eight out of ten are mothers, and two-thirds of them are living with a daughter. It is estimated that, when older family members are in need of constant care, 80% of such care is provided by kin, usually by wives and daughters.[17] It is interesting to note that the same clear trends, identifying women as carrying an extensive and complex load of family caring, have emerged in studies of welfare states. Although Norway eliminated family responsibility laws following the introduction of a "law for comprehensive care" that covered the young as well as the old, the care provided by Norwegian women has been described as "the hidden welfare state."[18]

It has been common, especially in the popular press, to suggest that women's involvement in the world of work will make them spend less time and effort on kin-tending. There is little evidence to support such a claim. Indeed, there are indications that an opposite trend is occurring. A growing number of women may be adjusting their work plans and work schedules to accommodate the needs of elderly parents[19]—much as they formerly planned around the needs of their children. Recent research found that employment significantly reduced caregiving to aging parents among sons, but this was not a statistically significant trend for daughters.[20]

There seems to be good reason to worry about what Betty Friedan has called "the superwoman squeeze"—the overload experienced by middle-generation women who provide support for both children and parents, in addition to facing the demands of the workaday world.[21] A growing number of writers express concern that our current social expectations regarding family giving help to the elderly are unrealistic—even dysfunctional, given recent demographic and social change. One asks: "At what point does the expectation of filial responsibility become social irresponsibility?"[22]

It is quite possible that, as a result of dramatic and rapid demographic change—particularly the enormous increase in the proportion of people who survive to advanced old age and face chronic health problems—we are finding that old attitudes and expectations about family care for impaired members simply do not work. The main casualties of this situation are likely to be middle-aged and young-old women, who face unmanageable care burdens or strong feelings of guilt. It is also important, however, to exercise some caution with regard to such conclusions. Because of the growing attention being devoted to the sick and the needy old, we may indeed go too far in equating "old" with "needy."

The Old as Resources

Television, newspapers, and popular magazines portray the old as a frail, dependent group that represents a drain on national and family resources. During the past year, a leading weekly news magazine has carried two cover stories focused on the elderly. The cover of the first read "Alzheimer's Disease: The New Epidemic." In the second, the bold-print title demanded to know "Who Is Taking Care of Our Parents?" Similarly, recent research and policy debates have tended to emphasize the neediness and dependency of the old, depicting them as a drain on scarce resources. Yet there is a striking lack of perception of the old as *constituting a resource,* a lack that ignores economic as well as psychological realities.

With regard to the family realm, few systematic efforts have been made to map the flow of material support from older generations to the young, even though it is clear that the family is by far the most important welfare or redistributional mechanism even in an advanced industrial country like the US.[23] Available research evidence indicates that, overall, the old in Western industrialized societies tend to give more economic assistance than they receive.[24]

We may have overlooked some critical "safety-valve" functions performed by older generations. When families make plans, for example, regarding major purchases, they often count on the older generation as a potential back-up if a crisis should intervene. Most older people own property, whereas it is becoming increasingly difficult for the young to do so. Often, the young may not actually end up turning to their elders, but their choices and behavior would have been more restrained if the older generation were not there as potential support. Such functions of the older generations may constitute an important and much neglected aspect of modern grandparenthood.

Evidence shows that grandparents may serve as indirect "stabilizers" of family life in a variety of ways. As a result of the clearer separation between parenthood and grandparenthood in the life of women, mothers may become more of a supportive force for their daughters' mothering.[25] Grandparents may help to render parents more understandable to their children, may function as arbitrators in conflicts between them, or may serve as confidants in difficult times. *How often* such mediation and arbitration occur may not be the right question to ask; rather, we might inquire as to what extent families perceive grandparents as a potential "safety valve." Nearly four decades ago, one writer described such functions of grandparents: "They stand ready to intervene as first and last aid as soon as the framework of the group is flagging or breaking up."[26] This observation was based on a decade when a world war had exerted tremendous stress on family units. Recently, other writers have pointed to two particular current social trends that are likely to activate grandparents as stress-buffers: the growing number of single adolescent parents and the high rates of divorce. In these cases, grandparents may take over some of the tasks of parenting. How often divorce leads to three-generational living, and how often grandparents provide substantial financial support to grandchildren following divorce, is not known, but a current study of divorce and grandparenthood shows grandparents on the "custodial side" to be significant factors in postdivorce adjustment.[27]

The Effects of Divorce and Remarriage

Many aging societies are, like the United States, divorcing societies. The historian Philippe Aries has suggested that this may be inevitable, because modern longevity makes marriage a much more long-term commitment than it was in the past. At the end of the nineteenth century, the average length of marriage until the time when one spouse died was about 28 years. In the late 1970s, it was over 43 years.[28] In the typical nineteenth-century family, one spouse was deceased before all the children were raised.

Nineteen hundred and seventy-four was the first year when more marriages in the US were terminated by divorce than they were by death.[29] Nearly half of recent marriages will end in divorce. Trends in divorce and remarriage are shaping the life courses and kin networks of men, women, and children, and each of these groups faces somewhat different challenges as a result of marital disruption and family reconstitution.

Studies of the differences between men's and women's divorce experiences paint a picture of severe financial losses for women, and of weakened family networks for men. There is little doubt that, for many women, divorce brings a severe reduction in standard of living and an uncertain financial future. This has ripple effects for children, who not only face reduced material resources, but who also have a mother who bears the strain of economic worries. In a study of college students whose parents had recently divorced, my colleagues and I found that nearly all said they worried about their mother's future. There is also good reason to think that divorce puts new stress on the parents of those divorcing, but they have tended to be the forgotten people in divorce research. For example, we do not know how often substantial financial aid is contributed to divorced children, especially daughters, nor do we know how often plans for retirement are postponed or otherwise altered as a result of children's marital break-ups. In cases of divorce after the age of 40, parents may find that their expectations regarding support from children, especially from daughters, are left unmet, because the divorce has depleted emotional and material resources. Like young adult children, such parents may also worry about the future of the middle-generation woman, who faces her own aging with uncertain support.

Among divorced women, roughly one-third never remarry. Overall, remarriage is becoming a disproportionately male experience in our society, especially after the age of 40. After the age of 65, remarriage rates for men are about eight times as high as for women.[30] This is in part owing to dramatically imbalanced sex ratios; in part, it reflects cultural norms regarding age differentials between spouses. Furthermore, research on remarriage following divorce in the early phase of adulthood has found that financial and educational resources operate quite differently in shaping remarriage probabilities among men and women. The more resources the woman has available (measured in education and income), the less likely she is to remarry.[31] For men, the trend is reversed. This suggests that the populations of those men and women who have divorced but not remarried are going to be quite different. Among women who divorced in the 1970s and 1980s, a number are resourceful individuals who already will have lived decades of life on their own as they face their early old age at the turn of the century.

There is little doubt that divorce and remarriage create more disruption in men's family networks than in women's. In a study of midlife divorce, my colleague and I found that many of the men expressed concern about the viability of family bonds. About a third felt that relationships with their children had suffered as a result of the break-up. When asked about their parents' reaction to the divorce, more than half of the men reported that their parents were worried about losing touch with their grandchildren. The corresponding figure for women was under 10%. In several parts of the interview, the men expressed a sense of "not having a family anymore." What some seemed to be struggling with was the loss of their kin-keeper, the wife.

As we noted earlier, recent studies suggest that divorce may lead to an intensification of bonds to grandparents on the "custodial side," but to a weakening of ties to the "noncustodial" grandparents. Based on findings from the midlife divorce study, it seems reasonable to conclude that custody is not the critical factor. Rather, contact between grandparents and grandchildren is typically mediated through the middle generation, even after the children are grown. Because the mother often serves as a kin-keeper for her husband's kin, divorce disrupts such mediation on the paternal side. The lack of contact between many divorced fathers and their children further aggravates this problem. Thus, a growing number of paternal grandparents face the loss of active grandparenthood as a ripple effect of their children's divorce.

A growing number of grandparents will also face the phenomenon of having step-grandchildren, although no figures are available on how common this experience is in our society. Much recent discussion has focused on how divorce and family reconstitution create complex kin relations for today's children. It has been argued, for example, that a growing number of children have more than two sets of grandparents. What we often forget is that marital disruption and remarriage have been part of childhood experiences throughout history; the only notable change is that the disruption is now commonly caused by divorce rather than death.[32]

Nevertheless, there are some ways in which the marital disruption of divorce creates complexities in kin relations that are not encountered when death is the cause of marital endings. Recent estimates state the probability of children under 16 experiencing the divorce of their parents at over one-third. Researchers provide some dramatic findings on the consequences of parental divorce for children's life courses. A large number of children lose virtually all contact with one parent, typically the father. Furthermore, among children who experience remarriage following divorce, nearly 40% will experience divorce number two.[33] The literature on divorce has paid little attention to the dissolution of step-relationships. A growing number of children, parents, and grandparents will have the experience of expending a great deal of effort on making step-relationships work, only to find them eventually dissolve. At the present time, such ex-relationships have no legal protection. Recent trends in divorce and remarriage give rise to a number of complex questions. How will children of divorce, remarriage, and redivorce approach the formation and maintenance of attachments during their own adult years? What patterns of support will exist between aging parents and children in families where the mother was divorced and turned to her children for help in earlier adulthood? What will be the relationships between aging fathers and children with whom they had only sporadic contact for many years, and for whom they did not pay regular support? Will the mother–daughter axis become even more important as the mainstay of family organization and cohesion in the twenty-first century?

Family Worlds and Life Experiences: A Widening Gap Between Men and Women?

Many of the changes explored in this chapter pertain more closely to life paths and family patterns among women than to those among men. Reduced mortality and greater longevity have been far more pronounced for women, and altered fertility patterns have reshaped women's adulthood more than men's. In addition, the more distinct demarcation between generations is more evident in female than in male lines, and the clearer sequencing of parenthood and grandparenthood is more typical of women's life course than of men's. The most durable intergenerational bond is that between mother and daughter, and more women than men experience life in multigenerational families. Such contrasts in demographic trends, combined with recent cultural changes, may indeed have created new or sharper differences between the social worlds of men and women, both in the family realm and in society at large.

Increased Matrilineality?

As we have noted briefly, women are kin-keepers, and the mother–daughter link is critical in the creation of family continuity and in the flow of intergenerational support. The historical shift from economic to emotional interdependence in the family, and the increasing role of nonfamily institutions in providing financial security for young and old, may have reduced core aspects of male family roles; however, they have further bolstered traditional female roles, such as caregiving and the provision of emotional support.*

Several reasons explain why families in the early twenty-first century are likely to have a stronger "female axis" than has recently been the case: the longer period of joint survival of mothers and daughters; historical changes in the role patterns of mothers and daughters; trends in divorce and remarriage; and nonmarital fertility. Mothers and daughters can now expect more than a half-century of joint survival. The first part of the next century may witness generations of mothers and daughters who not only share the experiences of growing old, but whose adult lives have shared more similarities than was the case in the latter part of this century. Women born in the 1940s belong to "transitional cohorts" between groups with strikingly different life experiences. It is not until the 1950 cohort that we see the emergence of "new women" who have high levels of education, smaller families, and fairly continuous work histories. These women will be young-old mothers of midadult daughters in the first decades of the twenty-first century.

If current trends in marital disruption persist, many mothers and daughters will spend a number of years when both are living

*Alice S. Rossi (1986) Sex and gender in an aging society. *Daedalus* **115,** 141–170.

without a spouse. Furthermore, as we noted earlier, divorce tends to weaken kin ties on the paternal side, although it may intensify kin relations in maternal lines. It has previously been estimated that, because of marital disruption, about 40% of children under 16 can expect to spend some time with only one parent. In over 90% of the cases, that parent is the mother. Often, such projections did not consider the possibility of nonmarital fertility. During the last few years, nearly 20% of all children born have had unmarried mothers. A current analysis that covers children born to married as well as to unmarried mothers provides some startling projections on the life experiences of children in the last two decades of this century.[34] The study estimates that, by the time children born in 1980 turn 17, more than 80% of them will have spent some time living with only one parent. Recently, writers have voiced concern not only about how recent social change has affected fathers and male family lines, but also about how it has strained the quality of men's anchoring in community life.

Contrasts Between Men's and Women's Social Networks

Alice Rossi* predicts that parenthood will be increasingly dissociated from marriage. She expresses concern that, when a growing number of mothers are solo parents, men are missing out on significant mechanisms to tie them into communal activities. Most women in our society experience a time when their identities in the surrounding community are defined in terms of being someone's mother. To what extent they go through a later phase when they are defined as someone's daughter, we do not know. What we do know is that, throughout their lives, women find it easier to cross age and generation boundaries than men do. Research has found that men, starting in boyhood, regard age differences as more salient and tend to orient themselves to horizontal relationships

*Alice S. Rossi (1986) Sex and gender in an aging society. *Daedalus* **115,** 141–170.

with age peers.[35] In the adult years, women's social networks are larger and more diverse than those of men.[36] For example, women are much more likely to have confidant relationships that are vertical—that is, to span generational lines. Recent trends in fertility and marital disruption may further weaken men's involvement in vertical ties, and lead to even greater contrasts between men's and women's networks.

Two demographers discuss men's declining involvement with young children and express concern about it as a trend.[37] They point out that between 1960 and 1980 the average male life course showed a dramatic decline in number of years involved with children, and they suggest that this may have serious consequences for the psychological development of men, such as their capability to be nurturant and altruistic, or for the quality of their attention to values and life priorities. The same authors also warn that men's declining involvement with children may have negative consequences for child welfare issues. They ask if men who do not have regular contact with children will be ready to support public spending on schools and social programs for the young.

All of this suggests that our society may currently be in a somewhat paradoxical situation. During recent decades, ideology has stressed equality between the sexes in their family roles. Yet, as we have seen, demographic and social changes have in many ways created very different family worlds for men and women. An increasing proportion of men have only precarious vertical ties, both up and down generational lines, whereas women's intergenerational ties are more varied, complex, and durable than ever before in human history.

Families are social arenas in which historical changes take on personal and shared meanings. They are also groups that meet critical human needs, and settings where biographies are written and rewritten as lives unfold, take on structure, and become interwoven. This chapter has reviewed some of the recent demographic and social changes that have transformed family life. Siblings, parents, and children, grandparents, and grandchildren now look forward to decades of shared biographies. Altered patterns of mortality have not only created relationships of unprecedented duration, but have also made the timing of family deaths more predictable. As the number of children per family has decreased, differences in life experiences among siblings have become reduced, and a greater proportion of family relationships are conducted across generational lines rather than with generational peers. Trends in fertility and mortality have resulted in increasingly "top-heavy" families, and family caregiving has become more and more focused on very old members. Multigenerational families have become more common, which means that a wider spectrum of kinship roles and relationships is open to family members. Finally, many of these recent changes have affected men and women quite differently, in some ways creating sharper contrasts between their family and social worlds.

Acknowledgment

Work on this chapter was supported by a Research Career Development Award from the National Institute on Aging, grant no. 1 K 04 AG 00203.

Notes and References

[1]R. Blythe (1979) *The View in Winter.* Harcourt, Brace, Jovanovich, New York, NY.

[2]For further discussion, *see* G. O. Hagestad and B. L. Neugarten (1985) Age and the life course, in *Handbook of Aging and the Social Sciences,* 2nd ed. (E. Shanas and R. H. Binstock, eds.), Van Nostrand and Reinhold, New York, NY, pp. 35–61.

[3]This discussion is based on the work by P. Uhlenberg (1980) Death and the family. *Journal of Family History* **5,** 313–320; and (1978) Changing configurations of the life course, in *Transitions: The Family and the Life Course in Historical Perspectives* (T. Hareven, ed.), Academic Press, New York, NY, pp. 65–97.

[4]H. H. Winsborough (1978) A demographic approach to the life cycle, in *Life Course: Integrative Theories and Exemplary Populations* (K. W. Back, ed.), Westview Press, Boulder, CO, pp. 65–76.

[5]T. K. Hareven (1977) Family time and historical time. *Daedalus* **106,** 57–70.

[6]A. S. Skolnick (1978) *The Intimate Environment,* 2nd ed. Little Brown, Boston, MA; John Demos (1970) *A Little Commonwealth: Family Life in Plymouth Colony.* Oxford University Press, New York, NY.

[7]Hareven, *Transitions: The Family and the Life Course in Historical Perspectives.*

[8]For a recent overview, *see* C. Feiring and M. Lewis (1984) Changing characteristics of the U.S. family, in *Beyond the Dyad* (M. Lewis, ed.), Plenum, New York, NY, pp. 59–89.

[9]S. H. Preston (1984) Children and the elderly in the U.S. *Scientific American* **Dec.,** 44–49.

[10]U. Lehr and W. Schneider (1983) "Fünf-Generationen-Familien: einige Daten über UrurgroBeltern in der Bundesrepublik Deutschland," *Zeitschrift für Gerontologie* **5,** 200–204.

[11]G. O. Hagestad (1985) Continuity and connectedness, in *Grandparenthood* (V. L. Bengston and J. F. Robertson, eds.), Sage Publications, Beverly Hills, CA, pp. 31–48.

[12]J. Modell, F. F. Furstenberg, Jr., and T. Hershberg (1976) Social change and transitions to adulthood in historical perspective. *Journal of Family* 7–32.

[13]M. White Riley (1983) The family in an aging society: A matrix of latent relationships. *Journal of Family Issues* 439–454.

[14]L. E. Troll, S. J. Miller, and R. C. Atchley (1979) *Families in Later Life* Wadsworth Publishing Co., Belmont, CA.

[15]E. Shanas and M. B. Sussman, eds. (1977) *Family, Bureaucracy, and the Elderly.* Duke University Press, Durham, NC.

[16]K. Morioka (1973) *Family Life Cycle: Theory Research and Practice.* Baifukan, Tokyo, Japan.

[17]Troll et al., *Families in Later Life.*

[18]K. Waerness (1978) The invisible welfare state: Women's work at home. *Acta Sociologica* **(suppl.),** 193–207.

[19]E. M. Brody (1979) Aged parents and aging children, in *Aging Parents* (P. K. Ragan, ed.), University of Southern California Press, Los Angeles, CA, pp. 267–288.

[20]E. Palo Stroller (1983) Parental caregiving by adult children. *Journal of Marriage and the Family* 851–858.

[21]B. Friedan (1981) *The Second Stage.* Summit Books, New York, NY.

[22]Brody, *Aging Parents* (P. K. Ragan, ed.), pp. 267–288.

[23]J. Morgan (1983) The redistribution of income by families and institutions in emergency help patterns, in *5000 American Families: Patterns of Economic Progress,* vol. 10 (G. Duncan and J. Morgan, eds.), Institute for Social Research, Ann Arbor, MI, pp. 1–59.

[24]R. Hill, N. Foote, J. Aldous, R. Carlson, and R. McDonald (1970) *Family Development in Three Generations.* Schenkman, Cambridge, MA.

[25]For a discussion of relevant work, *see* B. R. Tinsley and R. D. Parke (1984) Grandparents as support and socialization agents, in *Beyond the Dyad* (M. Lewis, ed.), Plenum, New York, NY, pp. 161–194.

[26]H. von Hentig (1946) The sociological function of grandmother. *Social Forces* 389.

[27]F. F. Furstenberg, Jr., J. L. Peterson, C. Winquist Nord, and N. Zill (1983) The life course of children of divorce: Marital disruption and parental contact, *American Sociological Review* **48,** 656–668.

[28]N. Goldman and G. Lord (1983) Sex differences in life cycle measures of widowhood. *Demography* 177–195.

[29]P. C. Glick (1980) Remarriage: Some recent changes and variations. *Journal of Family Issues* 455–478.

[30]National Center for Health Statistics: Advance report, final marriage statistics, 1980. *Monthly Vital Statistics Report,* vol. 32 , no. 4, supp. DHHS pub. no. (PHS) 83-1120. Public Health Service, Hyattsville, MD, Aug. 1983.

[31]G. B. Spanier and P. C. Glick (1980) Paths to remarriage. *Journal of Divorce* 283–298.

[32]J. Demos (1978) Old age in early New England, in *The American Family in Social-Historical Perspective,* 2nd ed. (M. Gordon, ed.), St. Martin's Press, New York, NY, pp. 220–256.

[33]Furstenberg, et al., *American Sociological Review,* 655–658.

[34]S. L. Hofferth (1985) Updating children's life course. *Journal of Marriage and the Family* 93–115.

[35]For a discussion of relevant work, *see* Hagestad and Neugarten (1985) *Handbook of Aging and the Social Sciences* (Shanas and Binstock), pp. 35–61.

[36]T. Antonnuci (1985) Personal characteristics, social support, and social behavior, in *Handbook of Aging and the Social Sciences* (Shanas and Binstock,), pp. 94–128.

[37]D. Eggebeen and P. Uhlenberg (1985) Changes in the organization of men's lives: 1960–1980. *Family Relations* 251–257.

What Do Grown Children Owe Their Parents?

Jane English

What do grown children owe their parents? I will contend that the answer is "nothing." Although I agree that there are many things that children *ought* to do for their parents, I will argue that it is inappropriate and misleading to describe them as things "owed." I will maintain that parents' voluntary sacrifices, rather than creating "debts" to be "repaid," tend to create love or "friendship." The duties of grown children are those of friends, and result from love between them and their parents, rather than being things owed in repayment for the parents' earlier sacrifices. Thus, I will oppose those philosophers who use the word "owe" whenever a duty or obligation exists. Although the "debt" metaphor is appropriate in some moral circumstances, my argument is that a love relationship is not such a case.

Misunderstandings about the proper relationship between parents and their grown children have resulted from reliance on the "owing" terminology. For instance, we hear parents complain, "You owe it to us to write home (keep up your piano playing, not adopt a hippie lifestyle), because of all we sacrificed for you (paying for piano lessons, sending you to college)." The child is sometimes even heard to reply, "I didn't ask to be born (to be given piano lessons, to be sent to college)." This inappropriate idiom of ordinary language tends to obscure, or even to undermine, the love that is the correct ground of filial obligation.

Aging and Ethics Ed.: N. Jecker ©1991 The Humana Press Inc.

Favors Create Debts

There are some cases, other than literal debts, in which talk of "owing," though metaphorical, is apt. New to the neighborhood, Max barely knows his neighbor, Nina, but he asks her if she will take in his mail while he is gone for a month's vacation. She agrees. If, subsequently, Nina asks Max to do the same for her, it seems that Max has a moral obligation to agree (greater than the one he would have had if Nina had not done the same for him), unless for some reason it would be a burden far out of proportion to the one Nina bore for him. I will call this a *favor:* when A, at B's request, bears some burden for B, then B incurs an obligation to reciprocate. Here, the metaphor of Max's "owing" Nina is appropriate. It is not literally a debt, of course, nor can Nina pass this IOU on to heirs, demand payment in the form of Max's taking out her garbage, or sue Max. Nonetheless, since Max ought to perform one act of a similar nature and amount of sacrifice in return, the term is suggestive. Once he reciprocates, the debt is "discharged"—that is, their obligations revert to the condition they were in before Max's initial request.

Contrast a situation in which Max simply goes on vacation and, to his surprise, finds upon his return that his neighbor has mowed his grass twice weekly in his absence. This is a voluntary sacrifice rather than a favor, and Max has no duty to reciprocate. It would be nice for him to volunteer to do so, but this would be supererogatory on his part. Rather than a favor, Nina's action is a friendly gesture. As a result, she might expect Max to chat over the back fence, help her catch her straying dog, or something similar—she might expect the development of a friendship, but Max would be chatting (or whatever) out of friendship, rather than in repayment for mown grass. If he did not return her gesture, she might feel rebuffed or miffed, but not unjustly treated or indignant, since Max has not failed to perform a duty. Talk of "owing" would be out of place in this case.

It is sometimes difficult to distinguish between favors and nonfavors, because friends tend to do favors for each other, and those who exchange favors tend to become friends, but one test is to ask how Max is motivated. Is it "to be nice to Nina" or "because she did *x* for me"? Favors are frequently performed by total strangers without any friendship developing. Nevertheless, a temporary obligation is created, even if the chance for repayment never arises. For instance, suppose that Oscar and Matilda, total strangers, are waiting in a long checkout line at the supermarket. Oscar, having forgotten the oregano, asks Matilda to watch his cart for a second. She does. If Matilda now asks Oscar to return the favor while she picks up some tomato sauce, he is obliged to agree. Even if she had not watched his cart, it would be inconsiderate of him to refuse, claiming he was too busy reading the magazines. He may have a duty to help others, but he would not "owe" it to her. However, if she has done the same for him, he incurs an additional obligation to help, and talk of "owing" is apt. It suggests an agreement to perform equal, reciprocal, canceling sacrifices.

The Duties of Friendship

The terms "owe" and "repay" are helpful in the case of favors, because the sameness of the amount of sacrifice on the two sides is important; the monetary metaphor suggests equal quantities of sacrifice. However, friendship ought to be characterized by *mutuality* rather than reciprocity: friends offer what they can give and accept what they need, without regard for the total amounts of benefits exchanged, and friends are motivated by love rather than by the prospect of repayment. Hence, talk of "owing" is singularly out of place in friendship.

For example, suppose Alfred takes Beatrice out for an expensive dinner and a movie. Beatrice incurs no obligation to "repay" him with a goodnight kiss or a return engagement. If Alfred complains that she "owes" him something, he is operating under the assumption that she should repay a favor, but on the contrary, his was a generous gesture done in the hopes of developing a friendship. We hope that he would not want her repayment in the form of sex or attention if this was done to discharge a debt rather than from friendship. Since, if Alfred is prone to reasoning in this way, Beatrice may well decline the invitation or request to pay for her own dinner, his attitude of expecting a "return" on his "investment" could hinder the development of a friendship. Beatrice should return the gesture only if she is motivated by friendship.

Another common misuse of the "owing" idiom occurs when the Smiths have dined at the Joneses' four times, but the Joneses at the Smiths' only once. People often say, "We owe them three dinners." This line of thinking may be appropriate between business acquaintances, but not between friends. After all, the Joneses invited the Smiths not in order to feed them or to be fed in turn, but because of the friendly contact presumably enjoyed by all on such occasions. If the Smiths do not feel friendship toward the Joneses, they can decline future invitations and not invite the Joneses; they owe them nothing. Of course, between friends of equal resources and needs, roughly equal sacrifices (though not necessarily roughly equal dinners) will typically occur. If the sacrifices are highly out of proportion to the resources, the relationship is closer to servility than to friendship.[1]

Another difference between favors and friendship is that, after a friendship ends, the duties of friendship end. The party that has sacrificed less owes the other nothing. For instance, suppose Elmer donated a pint of blood that his wife Doris needed during an operation. Years after their divorce, Elmer is in an accident and needs one pint of blood. His new wife, Cora, is also of the same blood type. It seems that Doris not only does not "owe" Elmer blood, but that she should actually refrain from coming forward if Cora has volunteered to donate. To insist on donating not only interferes with the newlyweds' friendship, but it belittles Doris and Elmer's former relationship by suggesting that Elmer gave blood in hopes of favors returned instead of simply out of love for Doris. It is one of the heart-rending features of divorce that it attends to quantity in a relationship previously characterized by mutuality. If Cora could not donate, Doris's obligation would be the same as that for any former spouse in need of blood; it is not increased by the fact that Elmer similarly aided her. It is affected by the degree to which they are still friends, which, in turn, may (or may not) have been influenced by Elmer's donation.

In short, unlike the debts created by favors, the duties of friendship do not require equal quantities of sacrifice. Performing equal sacrifices does not cancel the duties of friendship, as it does the debts of favors. Unrequested sacrifices do not themselves create debts, but friends have duties regardless of whether they requested or initiated the friendship. Those who perform favors may be motivated by mutual gain, whereas friends should be motivated by affection. These characteristics of the friendship relation are distorted by talk of "owing."

Parents and Children

The relationship between children and their parents should be one of friendship characterized by mutuality rather than one of reciprocal favors. The quantity of parental sacrifice is not relevant in determining what duties the grown child has. The medical assistance grown children ought to offer their ill mothers in old age depends on the mothers' need, not on whether they endured a difficult pregnancy, for example. Nor do one's duties to one's parents cease once an equal quantity of sacrifice has been performed, as the phrase "discharging a debt" may lead us to think.

Rather, what children ought to do for their parents (and parents for children) depends on (1) their respective needs, abilities, and resources and (2) the extent to which there is an ongoing friendship between them. Thus, regardless of the quantity of childhood sacrifices, an able, wealthy child has an obligation to help his or her needy parents more than does a needy child. To illustrate, suppose sisters Cecile and Dana are equally loved by their parents, even though Cecile was an easy child to care for and was seldom ill, whereas Dana was often sick and caused some trouble as a juvenile delinquent. As adults, Dana is a struggling artist living far away, whereas Cecile is a wealthy lawyer living nearby. When the parents need visits and financial aid, Cecile has an obligation to bear a higher proportion of these burdens than her sister. This results from her abilities, rather than from the quantities of sacrifice made by the parents earlier.

Sacrifices have an important causal role in creating an ongoing friendship, which may lead us to assume incorrectly that it is the sacrifices that are the source of the obligation. That the source is the friendship instead can be seen by examining cases in which the sacrifices occurred, but the friendship, for some reason, did not develop or persist. For example, if a woman gives up her newborn child for adoption, and if no feelings of love ever develop on either side, it seems that the grown child does not have an obligation to "repay" her for her sacrifices in pregnancy. For that matter, if the adopted child has an unimpaired love relationship with the adoptive parents, he or she has the same obligations to help them as a natural child would have.

The filial obligations of grown children are a result of friendship, rather than owed for services rendered. Suppose that Vance married Lola despite his parents' strong wish that he marry within their religion, and that as a result, the parents refuse to speak to him again. As the years pass, the parents are unaware of Vance's problems, his accomplishments, and the birth of his children. The love that once existed between them, let us suppose, has been completely destroyed by this event and 30 years of desuetude. At this point, it seems, Vance is under no obligation to pay his parents' medical bills in their old age, beyond his general duty to help those in need. An additional, filial obligation would only arise from whatever love he may still feel for them. It would be irrelevant for his parents to argue, "But look how much we sacrificed for you when you were young," for that sacrifice was not a favor, but occurred as part of a friendship that existed at that time but is now, we have supposed, defunct. A more appropriate message would be, "We still love you, and we would like to renew our friendship."

I hope this helps to set the question of what children ought to do for their parents in a new light. The parental argument, "You ought to do *x* because we did *y* for you," should be replaced by, "We love you, and you will be happier if you do *x,*" or "We believe you love us, and anyone who loved us would do *x.*" If the parents' sacrifice had been a favor, the child's reply, "I never asked you to do *y* for me," would have been relevant; to the revised parental remarks, this reply is clearly irrelevant. The child can either do *x* or dispute one of the parents' claims: by showing that a love relationship does not exist, or that love for someone does not motivate doing *x,* or that he or she will not be happier doing *x.*

Seen in this light, parental requests for children to write home, visit, and offer them a reasonable amount of emotional and financial support in life's crises are well founded, so long as a friendship still exists. Love for others does call for caring about and caring for them. Some other parental requests, such as for more sweeping changes in the child's lifestyle or life goals, can be seen to be insupportable, once we shift the justification from debts owed to love. The terminology of favors suggests the reasoning, "Since we paid for your college education, you owe it to us to make a career of engineering, rather than becoming a rock musician." This tends to alienate affection even further, since the tuition payments are depicted as investments for a return rather than done from love, as though the child's life goals could be "bought." Basing the argument on love leads to different reasoning patterns. The suppressed premise, "If A loves B, then A follows B's wishes as to A's lifelong career" is simply false. Love does not even dictate that the child adopt the parents' values as to the desirability of alternative life goals. So the parents' strongest available argument here is, "We love you, we are deeply concerned about your happiness, and in the long run you will be happier as an engineer." This makes it clear that an empirical claim is really the subject of the debate.

The function of these examples is to draw out our considered judgments as to the proper relation between parents and their grown children, and to show how poorly they fit the model of favors. What is relevant is the ongoing friendship that exists between parents and children. Although that relationship developed partly as a result of parental sacrifices for the child, the duties that grown children have to their parents result from the friendship rather than from the sacrifices. The idiom of owing favors to one's parents can actually be destructive if it undermines the role of mutuality and leads us to think in terms of quantitative reciprocal favors.

Reference

[1]Cf T. E. Hill, Jr. (1973) Servility and self-respect. *Monist* **57,** 87–104. Thus, during childhood, most of the sacrifices will come from the parents, since they have most of the resources, and the child has most of the needs. When children are grown, the situation is usually reversed.

Families as Caregivers

The Limits of Morality

Daniel Callahan

How are we to understand and live our lives when the moral demands made upon us require more than we can give, more than we can make any sense of, and—in our society at least—more than commands much respect and admiration? I raise this disturbing question in the context of a developing trend to return to families and the home the long-term care of the chronically ill and those in need of rehabilitation.[1] The assumption behind the promotion of this trend is that families will, with some modest degree of social support, be able practically to manage such care, and have the moral, psychological, and spiritual strength to do so.

However, is that true? Family care can be a mutually rewarding experience for those who are cared for and those who do the caring, a source of growth and mutual enrichment. The demands it makes can be strenuous and of a kind requiring a new self-understanding for both patient and family; and that can be the occasion of fresh strength and moral achievement.[2] However, it can also be an occasion of oppression and hostility; the caregiver is often trapped in a way of life not chosen and a future direction not of his or her own. Our secular morality (although perhaps not our religious traditions) provides few resources for living lives of unchosen obligations, those that through mischance lay upon us overwhelming demands to give our life over to the happiness and welfare of someone else.

Aging and Ethics Ed.: N. Jecker

The idea of a fresh emphasis on family support is both ironic and appealing. Part of the ideology behind medical and technological progress is that of freeing human life from the inexorability of bodily decay and disability and, at the same time, from the uninvited and smothering social burdens they impose on our individual and communal life. It is then ironic that that progress should lead us back to embracing just those same burdens.

The more understandable and seductive impetus behind this trend is that of the financial pressure occasioned by the growing social burden of disability. An ever-growing proportion of people are kept alive for an ever-longer period of time—but at the price of ever-extended care and rehabilitation to insure their continuing survival and well being. The full provision of such care by government funds or institutions promises to be insupportable. Too many people need too much care. What is to be done? The popular answer is to widen the scope and acceptability of family care, as a less expensive solution. A supporting motive is a widespread belief that family care is superior care, more kindly and sensitive, more acutely attuned to the needs of individuals, and more compatible with traditional values, notably those of kinship and family integrity.

What kind of moral resources are necessary to carry out such a policy? On the basis of what ethical foundations can society ask people to take on the often heavy, sometimes overwhelming, burden of care for another? Even if people are willing in principle to care for a fellow family member, where can they personally find the necessary moral resources to sustain their commitment, to make moral sense of it? If they can solve that problem, how will society honor and help them, and how will it provide a social meaning to complement and reinforce whatever individual meaning they may bring to bear?

A basic premise of medical progress, and the modernism of which it is a part, is that we need not bow down to the raw deliverances of nature; and one of those deliverances is the way illness makes us a burden on each other, needy and dependent. Medicine has sought to find a means of liberation from that affliction no less than from the more direct afflictions of the body, but how far are we justified in going to escape the burden that the illness of another can impose on us, and does the fact that the illness may go on much longer than in the past make any difference in thinking about our moral obligations?

It is difficult to find an adequate answer to the question of what we owe each other in times of stress and crisis, and all the more so when the demands made on us seem to threaten deeply our own happiness and fulfillment. We moralize readily enough about keeping our promises, honoring our commitments, and respecting our contracts, but if doing so begins to threaten our psychological survival, our basic social freedom, our otherwise legitimate private hopes and plans, then only the most well rooted, the most cogent reasons, and compelling emotions are likely to sustain us. That goal is not easy in our society. We lack as a people common vision of the wellsprings of moral obligation, a shared understanding of the moral significance of pain and suffering, and a clear notion of how we ought to support each other's private griefs and burdens. Nor do we strongly encourage those personal virtues and character traits that enable people to endure in the face of adversity.

The instinct for survival is well imbedded in modern morality.[3] Social morality has become cagey, ready enough to invoke firm rules and principles. Racism, sexism, religious or ethnic discrimination, we well know, admit to no moral wheeling and dealing, no compromise or accommodation. Matters are otherwise with personal morality, where autonomy, choice, survival, and realistic coping are honored—and therapeutic sages well paid to help us escape being trapped in moral corners. After all, what is the point of life if one cannot pursue happiness and have some chance of finding it?

Moreover, quite apart from cultural support, it is an old and hard moral question to know what we should make of demands for self-sacrifice. Most moral rules have common sense and practicality to commend them. Murder, lying, and theft ordinarily have tangibly bad consequences for those who commit such acts. Even our self-interest commends us to avoid them. Matters are otherwise when we are morally asked to give up our lives, or personal hopes, for the sake of another. Only under special circumstances can that seem to make any sense at all from the viewpoint of self-interest, even of the most benign sort. It is not for nothing that almost all Western moralities have been careful to distinguish between duty and supererogation.[4] They all recognize that a morality designed to apply to a whole community cannot require that everyone be a saint or a selfless paragon of altruism. The notion that we might as a matter of social policy burden families with the heroic duty of caring for the chronically ill or those in need of a course of rehabilitation that may fail and render them chronically ill or disabled ought at least to raise a red flag of warning.

What can we realistically ask of people? What can they realistically ask of themselves? If we have an obligation to care for each other as family members when the need arises, what are the limits of that obligation? If it is perhaps reasonable to demand by means of persuasion and social pressure that we discharge our obligations toward our family members, it may not always be equally reasonable to give those obligations legal sanction. Indeed, wisdom might suggest that we make such obligations morally attractive rather than coercive, if only for the sake of those fated to receive the care (who may be presumed to prefer care graciously, rather than grudgingly, given). We need to discover not only what might be morally required, but where the borderline of duty ends. We need also to discover, if possible, how we are to make that duty emotionally satisfying; that is, how to make it productive of meaning and coherence, not of merely arid self-denial and the austere pleasure of doing one's duty for its own sake.

That is by no means an easy task, and it is helpful to look at the realities of family care to understand just how problematic it can be in some circumstances given our present moral resources.[5] The problem itself is not easy to talk about; it invites undue optimism or undue pessimism. That many families do exceedingly well cannot be denied; both the family and the patient find new personal resources and adapt well. Far from being simply a burden, the patient can often make a significant contribution to the family providing the care. That frequent outcome, however, invites an overly optimistic, sometimes rationalizing, attitude; there can be a subtle implication that those who fail to flourish do so because of some character flaw, but it is possible both to recognize that some families and patients do exceedingly well, whereas others, through no apparent fault, find the situation more than they can bear.

I want to put to one side, for the moment, all those aspects of care that can be satisfying and those circumstances where it works out well enough. I want to focus, instead, on those features of care that impose the sharpest moral demands on the caregiver, in essence, those features that seem to pose a direct and fundamental threat to the welfare and happiness of the person who gives the care, where the caregiver may become—by the sacrifice demanded and extracted—as much a victim of the illness or disability as the person who is cared for.

How might we best think of the moral situation of a family member called on to provide care? Different diseases and conditions modify any general answer, as must different social and familial circumstances, but a common thread is that of a person unfreely and unexpectedly called on to provide a level of care well beyond that ordinarily demanded in family life. I use the term "unfreely" to indicate a lack of initial choice in providing the care. One is drafted by circumstances, and sometimes just as roughly and abruptly as sailors used to be impressed into the British navy. It happens "unexpectedly" in the sense that the family relationship that imposes the moral demand was not originally envisioned as one that would make a radical demand on the self; one feels the victim of capricious and inexplicable bad luck. To marry, as a common vow has it, "for better or for worse, in sickness and in health" suggests that life together may not turn out well; but few actually expect the worst to be realized in marriage. They might well have made another choice had they thought it at all likely. The "worse," then, usually comes as a shock and surprise. That in some cases the person requiring the care is, because of damaging changes wrought by the precipitating accident or illness, no longer the same person as before can only intensify the pain. One is being asked to give of oneself in ways that would otherwise have been unimaginable, and unimaginably unacceptable had it been possible to spell them out in advance.

How do people react to that combination of circumstances? Clearly, there are differences between parents caring for a handicapped child, a young wife caring for a husband recovering from a spinal cord injury, an elderly husband dealing with the hemodialysis of his wife, or a middle-aged wife helping to rehabilitate a husband struck down by a stroke. Nonetheless, just as those who fall in love, lose a job, or try to cope with the grief of a death will go through many similar feelings even in quite different circumstances, so too there seems to be a striking similarity to the response of caregivers. To be trapped as a caregiver, forced to empty oneself in what is often a one-sided relationship, can be an unendurably great burden. Let me try to sketch a composite portrait of family caregivers (keeping in mind that I am stressing the burdens, not the possible satisfactions).[6–10]

They are likely, most commonly, to feel anger that an unwanted fate has been visited upon them, an anger as often turned inward as toward the person who is ill. Why has this happened to *me?* Will I ever escape? Can anyone possibly understand what this costs me? The anger, in turn, often generates guilt: sometimes because the anger is aggressively turned toward the ill person—one who may be perfectly blameless but is, nonetheless, the cause of the problem; and sometimes because one feels one has failed oneself, failed by virtue of one's otherwise hidden anger and rebellion to live up to moral ideals, to marital commitments, or to what a parent is supposed to owe a child. That others do not notice the failure is beside the point; one's conscience knows and that is enough. One is, at once, on the outside a noble and giving person, gamely and lovingly facing up to adversity, and, on the inside, rebelling with hostility at the self-giving that is unfairly, even outrageously, demanded. Despite the fact of providing care, the sense of self-worth is severely compromised. Some who suffer in silence and the privacy of their hearts come to fear that they will be found out, that as others come to know them better they will be exposed for what they are, not the loving and long-suffering person that appears on the surface, but a petty, self-centered, self-pitying person, full of anger and hostility.

Anger and guilt play on each other, tearing at one's self-image and gnawing away at the bond between the caregiver and the ill person. With great effort, the grosser forms of anger and rebellion can often be brought to heel, only to issue in the slow torture of suffocating irritation and barely repressed irascibility. One lives life on the edge of losing one's temper and one's self-respect.

That such a combination of feelings should on occasion produce fantasies about the death of the ill person is hardly surprising. It is the perfect imaginative solution to the unwanted moral burden of caring for another, at once decisively final in the liberation it promises and utterly acceptable as a social solution. No more ideal a resolution can offer itself to the conscience, but until it happens—and it may not happen for years—the conscience is all the more burdened; it seems almost a form of murder to have such thoughts, but how could one not have them?

If the person being cared for were a stranger, matters might be otherwise; some distance would be possible, some saving detachment. In the case of family burdens, what is being drawn on and used as a moral noose is the very love and affection that are supposed to be the mark of such relationships. It is a gross distortion of a bond that is supposed to give us comfort in life. Instead, as a kind of cruel *reductio ad absurdum,* that bond is turned into a yoke that suffocates, making one pay many times over for having hoped for love and joy and having, at one time, perhaps achieved it.

What can it be for a mother—joyous upon the delivery of a child who turns out to be handicapped and requiring great care—to find that she will not have the child she wanted, or the kind of motherhood that is among the more benign of life's aspirations or the future that had played in her imagination as she waited for her child's birth?[11] What can it be for the young wife of a stroke victim to find that she must change her ideas of what marriage should be, her ideas of what the social scientists all too coolly call "role expectations?"[12] In some cases, the mutuality that was meant to mark their relationship, and at one time actually did so, can barely be recaptured, if at all. Their circle of friends may diminish if her husband cannot negotiate the ordinary demands of a social life; thus, to all the other burdens is added that of social isolation. Stigma is sometimes present.

Finally, there is the question of prognosis. Uncertainty of prognosis seems one of the great burdens of rehabilitative care. Will it succeed or will it turn into chronic care? Has the caregiver been sentenced to life imprisonment, or to a short term only? Not knowing the answer to that question can lead to an emotional seesaw of grand proportions, the caregiver tempted one day to see progress in the faintest signs of improvement and no less tempted the next to see no evidence of movement whatever.

At stake here is the possibility of hope. All things may be endurable if the demands are finite in depth and time, but a future that offers no exit at all, even if the burden on a daily basis is not utterly overwhelming, can be an obvious source of sadness and depression. Time is a curious phenomenon in that respect. Our past is always over and gone, and the present transitory and ephemeral. Although the wise person is probably someone who knows how to live in and make the most of that transitory present, if the present is unrelenting in its demands, all we have then is our future. No burden can be greater than trying to imagine how one can cope with a future that promises no relief. That is the very meaning of despair and why it is the ultimate human misery. If time is (so it is said) the cure for unrequited love, it might well seem a curse for that requited love that binds us forever to someone as a caregiver.

I have dwelled on some of the details of the moral demands that can be imposed on caregivers, because I think it important to understand what is being required of them. Put in the starkest terms, they are being asked by circumstances to sacrifice their selfhood and their own futures for the sake of another, and for some undeterminable and possibly interminable period. Also, they are doing so (at least initially) not because they chose such a fate, but because it was cast on them by virtue of the bonds of love or duty or both. Because at an earlier time they ventured, by a voluntary act, to give of themselves to another, they are now asked to live out some horrible implications of that choice, however unexpected and unwanted they may be. If they do not accept that burden, morality seems to say, they will betray their earlier commitment, and with it, the lives of the people who trusted them. If, however, they take up the commitment in its full rigor, they may do so at the cost of their own lives and futures. Is that fair, and can so much be asked of us?

In trying to deal with that last question, I want first to note what I mean here by moral obligation, and particularly the kind and basis of obligation that seems to present itself in these cases. By moral obligation I mean a justifiable claim made on us to act on behalf of the welfare of another, whether or not so acting is convenient or gratifying; only a higher, more encompassing moral obligation can displace a moral obligation, not claims of unhappiness or inconvenience.

What is the basis of moral obligation in the case of family relationships? It is initially tempting to make many distinctions here. Does it not seem intuitively obvious that there must be some fundamental differences between what a parent owes a child, what a wife owes a husband, what a brother owes a sister, and what a child owes an elderly parent? Should one not distinguish between the kind of contractual relationship that originated a marital bond, and the biological origins of a parent–child relationship?

Without trying to present a full argument, I want to suggest that the particular nature of the family bond is not nearly so important as a special feature of the relationship. The feature I have in mind is the ultimate neediness and vulnerability of the family member who requires care.[13] The perfect and classical model of such neediness is the dependence of an infant on its parents; its vulnerability is so complete that it cannot even exist unless that is recognized and responded to, but adverse circumstances can make others almost as vulnerable and helpless. It is not necessarily that someone else could not, in principle, provide the care, just as an anonymous wet nurse can save the life of a hungry child, or a series of hired, indifferent nurses care for the physical needs of an elderly person.

Instead, the important aspect of the vulnerability of ill or injured family members is that they may want and need the kind of care that only someone close to them, an integral part of their lives, can provide. A wife may need her husband, not just any person, to care for her; she is the one who once chose to cast her lot with him and took the trouble over the years to come to know him in a way that is not likely for another. More important, there is no reason why another should care about her welfare as much. What vulnerability most requires is someone who deeply cares, someone who will remain faithful—but faithful to us as a special and distinctive person, not as a mere object of moral duty or universal love. All this is simply to say that at the heart of any significant moral obligation is the vulnerability of another and, in the context of family life and illness, a vulnerability that can only adequately be responded to by a family member. Many exceptions and quibbles can be imagined to qualify that general assertion, but they are not important to my main point.

That point brings me to the heart of the moral problem. If we grant that one family member can desperately need the care of another, and *only* that other can respond to that need in a fully adequate way, does that automatically entail a right (explicit or implicit) on the part of the sick or disabled person to such care, and a corresponding obligation of the family member to provide it? It might at once be responded that, even if they initially resist being cared for by those who are not family members, patients can and usually will adapt well enough, or it might be noted that many of those in need of rehabilitation would prefer to have others provide their physical and health care. I do not want to deny those possibilities, but only want to focus more sharply on those situations where the demands on family members are not so readily relieved.

I will assume that there is considerable agreement that, if the demands are not great or excessive, we do assume that existence of some strong family obligations: parents ought to care for their ill children, husbands for their ill wives, and children for their elderly ill parents. Minimal decency seems to require that.[14] However, the idea of unlimited self-sacrifice encounters heavy, and perhaps increasing, resistance. One line of objection is primarily practical; that if it is true that "ought" implies "can," then the placing of excessively heavy burdens on people is simply unwise. They will collapse under the pressure, and perhaps in the process simply increase the problems of those for whom they are supposed to care. More broadly, it could be argued that it would be foolish to advance family care as social policy if the net result would be a sharp increase in divorce, physical and mental abuse, widespread neglect, and other evidence of demands too severe to be widely borne.

I am, however, more interested in noting another line of objection, one that focuses on the morally reasonable demands of self-love and self-interest as a way of establishing the limits of obligation.[4] Contemporary moral philosophers, for instance, show considerable nervousness about extending the scope of moral obligation much beyond explicitly understood and accepted contractual agreements. To count as moral at all, an action must stem from a free choice; autonomy is an underlying requirement. The idea of noncontractual moral obligations thus becomes highly problematic. They violate our autonomy, and they fall into the realm of supererogation, commendable, virtuous, and edifying if we freely choose them, but not required in the name of morality. The purpose of drawing such a sharp line is not necessarily to aggrandize the self or lead it into the green pastures of unfettered self-interest. Instead, it stems in great part from the fundamental difficulty philosophers have had in establishing a solid moral basis for involuntary self-sacrifice springing from contextual, not contractual demands. That difficulty stems, I believe, from the essentially individualistic orientation of much contemporary Anglo-American moral philosophy.

Two other intellectual streams offer somewhat different but converging approaches. Feminism has been particularly concerned with combating a culturally reinforced trait of women to all-too-readily embrace self-sacrifice and a selfless life. Since most family caregivers are women, that critique has special importance.[15] From another point of departure, Protestant theology since World War II has tried to find a better fit between the central Christian virtue of love of others and the more modern insights into the value and necessity of love of self.[8] In particular, it had to take apart the idea that love must, of its nature, be selfless, wholly other-directed. How, many come to ask, can one love another with openness and integrity unless one also loves oneself in some significant way? Can a person who does not love himself or herself love another?

It is far easier to discuss the general problems of moral obligation than to offer specific guidance on setting limits. The most obvious reason is that it is difficult to generalize about the capacity of people to take on heavy moral burdens. Not all people seem equally able to give to others or to be able to affect greatly their emotional capacities to do so, even with the best will in the world. We cannot, therefore, readily establish any set of reasonable expectations of caregivers when heroic, extraordinary caregiving is needed. That kind of caregiving pushes beyond the ordinary bounds of morality.

However, that reality suggests a deeper issue about morality itself. Morality in general—but self-sacrificial morality in particular—cannot be sustained by will alone. A presumption of much secular, individualistic morality is that one ought to do one's duty because it is one's duty, and that good reasons for moral behavior are sufficient motivation. One need only make up one's mind to act in a certain way, will to so act, and the actions will follow. But there are too many good-willed but still angry caregivers around to sustain that view. It is psychologically naive. It takes account neither of our emotions (which color our judgment and will from the inside) nor of the social setting of our actions (which influences our judgment and emotions from the outside).

With minor moral rules and moral demands, will alone may sustain us; and, as noted, the following of most moral rules has practical benefits—we get as much as we give, but when we are asked to make sacrifices that seem to promise far more burdens than benefits, much less the more extreme situation of forfeiting a life of our own choosing and direction, then we move beyond what calculating reason can justify or all but the stoutest of wills can will. At that point, we are forced to ask how our moral actions give meaning to our life. If the keeping of a moral rule, or the imposition of a heroic demand, seems to threaten our otherwise perfectly legitimate claim to happiness, then some deeper justification than morality alone seems needed. How can we, willingly, be asked to embrace personal tragedy in the name of morality alone? Why should I give up my own happiness for the sake of the care needed by another? There is no good general answer to that question within the compass of a morality of rules, duty, and a will to do the right thing. We can praise those who make great sacrifices, but we cannot readily condemn those who do not. We seem to be at a moral impasse.

Is there a solution? Heroic self-sacrifice, I have come to think, is only possible if understood within the context of an entire way of life, and a way of life set ultimately within some scheme of religious or higher meaning. I suggested above that it is the vulnerability of others that is the source of their claim on us, the fact that we and only we can provide the care they need, a care that responds to them as unique individuals and not merely needy examples of *homo sapiens*. That we are called on to respond to someone else's vulnerability means that, in turn, we become vulnerable also. Something is going to be taken from us as the ineluctable price of self-giving, perhaps something as central as our hopes and our identities.

I can conceive of myself making a radical sacrifice for another if I live in a community that understands the interrelationship of all our mutual needs and vulnerabilities and creates a society to respond to them. It is precisely because life so often fails our expectations that we need each other. It is precisely because the moral demands to give of self can be so outrageous, so utterly devastating, that we need to know others are prepared to do the same for us. It is the isolation of the moral claim to heroic action that is so intimidating in our kind of society. We cannot be sure that others will sustain us, or that others would do likewise for us, if our needs become heavy. Why should we expect any such things if all such sacrifices are thought to be heroic, and thus utterly optional?

In one sense, what I am saying points in the direction of improved systems of social support for those who care for family members. They need the financial and psychological support of state and federal agencies, and they need responsive, sensitive people to give them help and to give them respite, but that is hardly enough. In another sense, however, we need a different kind of society and a different kind of morality. Even with adequate social support, we are still faced with tragedy, still faced with moral claims that seem to confront us with imperative duties that are, for all that, impossible demands. How are we to give meaning to those demands? Morality alone cannot do so, because morality does not give meaning to life; it cannot lift itself by its own bootstraps. Part of the problem here turns on whether we take the harm done by those injuries and illnesses that lie behind the need for rehabilitation to be that of the injuries themselves or only our social response to them. The aggressive rationalistic and scientific tradition in medicine would look on the injuries themselves as an evil to be eradicated. Another interpretation, however, is that injuries are just chance occurrences of nature; any evil lies exclusively in the failure of the community to respond to the needs of those afflicted by them. Whichever ultimate interpretation we may prefer, however, there will still be for those families that must provide care a sense of personal tragedy. Something terrible has happened to them, and some meaning must be given to it.

On the whole, it seems to be religious cultures alone that can provide the kind of meaning needed, at least so far in human history. Suffering must be understood to have a point and to be redeemable. The care of another must be transformed from a stark and unpalatable moral demand to a satisfying moral vocation, one honored by the community and returned in kind when the caregiver comes, as we all will, to need care. Vulnerability is understood to be part of the human condition, some being visited with greater needs than others, but others being blessed with greater strength. A distinction is still made between duty and moral heroism, but the level of duty is set much higher, there is less nervousness about the boundary line, and heroism is thought to be required of all lives from time to time. A good society is one that finds ways to match needs and strengths, one that cares not only for public injustice visited on minority and other weak groups, but also for private injustices that nature and life visit upon individual people.

I said that I think only religious cultures have been able to project, and sometimes embody, that vision of community. For many of us who are not religious believers, we are left with a severe problem. How can we create a secular version of a way of life that fully shares burdens? I am not certain, but until we do I think we should be wary of asking families to undertake heroic sacrifices. If we must ask them, then it becomes imperative that we find out how we might best reward and sustain them. It seems to me possible that we might find ways to do that.

Notes and References

[1]J. P. Ackford (1979) Reducing medicaid expenditures through family responsibility: Critique of recent proposal. *American Journal of Law and Medicine* **5,** 59–79.

[2]E. M. Brody (1985) "Women in the middle" and family help to older people. *Gerontologist* **21,** 471–480.

[3]M. Mailick (1979) Impact of severe illness on individual and family: Overview. *Social Work in Health Care* **5,** 117–128.

[4]J. O. Urmson (1958) Saints and heroes, in *Essays in Moral Philosophy* (A. I. Melden, ed.), University of Washington Press, Seattle, WA, pp. 198–216.

[5]W. H. Jarrett (1985) Caregiving within kinship systems: Is affection really necessary? *The Gerontologist* **25,** 5–10.

[6]A. Golodetz (1969) Care of chronic illness: "Responsor" Role. *Medical Care* **7,** 385–394.

[7]A. MacIntyre (1983) *After Virtue.* University of Notre Dame Press, Notre Dame, IN.

[8]G. Outka (1972) *Agape: An Ethical Analysis.* Yale University Press, New Haven, CT.

[9]M. Skelton (1973) Psychological stress in wives of patients with myocardial infarction. *British Medical Journal* **2,** 101–103.

[10]J. M. Warrington (1981) *The Humpty Dumpty Syndrome.* Light and Life Press, Winona, IN.

[11]B. Holadays (1984) Challenges of rearing chronically ill child. *Nursing Clinics of North America* **19,** 361–368.

[12]J. O. Carpenter (1974) Changing roles and disagreement in families with disabled husbands. *Archives of Physical Medicine and Rehabilitation* **55,** 272–274.

[13]R. E. Goodin (1985) *Protecting the Vulnerable: A Reanalysis of Our Social Responsibilities.* University of Chicago Press, Chicago, IL.

[14]J. Bluestein (1982) *Parents and Children: The Ethics of the Family.* Oxford University Press, New York, NY.

[15]E. M. Brody (1985) Parent care as normative family stress. *Gerontologist* **25,** 19–29.

Health Care and Decision Making

Sara T. Fry

The status of the elderly in American society has changed remarkably within the last fifty years. People live longer and they appear to be in better health than their elders in earlier years.[1] The elderly also comprise a larger proportion of the population than ever before.[2]

Yet the joys of living longer and living well are not really enjoyed by many of our senior citizens. Some elderly individuals are simply not in good health. They are often called the "frail elderly," are dependent on others, and seem to require a larger proportion of health care resources and social services during their final years than ever before.[3] Other elderly do not have immediate family or relatives who provide housing, care, or financial or emotional support.[4] They live alone or in institutions that seem designed to meet caretaker needs rather than the needs of the elderly themselves.[5] A significant portion of the elderly even live in poverty or near-poverty conditions and do not have the capacity to change their situations.[6] Sadly, many elderly citizens have given up on life in an affluent society as evidenced by the growing number of elder suicides in recent years.[7,8]

Aging and Ethics Ed.: N. Jecker ©1991 The Humana Press Inc.

Issues for Health Professionals and Family Members

For health professionals and family members, these changes in the status, health, and welfare of the elderly create a number of social and moral issues. Who is responsible for the elderly? The state? Local communities? Their children or relatives? How much and what kind of public and private resources ought to be set aside for the care of the elderly? Should this care be directed toward care for acute illness or should it be earmarked for long-term care or nursing care supervision of the elderly at home? Is there a limit to how much health care and what types of health care resources the elderly should receive? Who should make this decision? Should the elderly simply be allowed to choose for themselves? When they can no longer make choices because of decreasing physical and mental health, who should choose for them? Their family members or health professionals who are presumed to have their best interests at heart?

How nurses and families, as opposed to physicians or patients, confront these issues in the daily care of elderly individuals in home-based care, long-term care, and acute care is the subject of this chapter. At the heart of all of these issues is the extent of obligations to respect the self-determined choices of elderly citizens and to provide benefit to them. A related issue is the limitations of elderly autonomy that might reasonably be imposed under certain conditions.

Respecting Choice While Providing Benefit

No one denies that all individuals, even the elderly, should have the opportunity to enjoy as much autonomy as possible within our society. Autonomy, a treasured moral value of our liberal society, is even advocated for the elderly in much the same way that autonomy is advocated for everyone. More housing options, greater health resources, and legislated public assistance programs are all ways that we tend to advocate autonomy for elderly in contemporary society.[9] Autonomy might even be considered a right that may be claimed by the elderly, and respect for autonomy is a duty that ought to be fulfilled, especially in the health care arena.

However, how much autonomy is really possible where the majority of elderly individuals are concerned? How can one be autonomous when hampered by decreasing physical and mental abilities and decreased social support networks? After all, to realize autonomy one must be a self-determining agent capable of acting on the plans and rational choices that one makes. The reality of what it means to be elderly indicates that the amount of autonomy that an elderly person enjoys is extremely limited with the passing of years. This is a fact that affects lifestyle, happiness, and even how and when one dies.

To say that health professionals or families ought to respect autonomy is to assert a principle of autonomy as a guideline for action on the part of the professional.[10] Autonomy, so conceived, is a *prima facie* principle indicating that it is to be followed unless it can be overridden by another moral principle of greater weight or standing. Autonomy, however, can also be understood as an ideal and used as the basis for arguing against something that attempts to impose a set of ends, values, or attitudes on an individual.[11] Thus, it is a capacity exercised by individuals when they examine their preferences and values, and that gives meaning and coherence to their lives.[12]

Both the ideal and principle of autonomy are woven together in health care decision making for the elderly. The principle may be hard to operationalize in situations where autonomy is compromised by illness, decreasing mental and physical abilities, and institutional rules and regulations. In such situations, even the ideal of autonomy may be lost when the capacity to make autonomous choices is constrained or judged nonexistent.

Limitations of Elder Autonomy

Constraints on elder autonomy are generally manifested in two ways. One way is through internal constraints, which may reduce or eliminate the capacity of the elderly person to claim autonomy. A second way is through external constraints, which allow legal, moral, and institutional overriding of autonomy for reasons perceived to be in the best interests of the individual.[13]

Internal Constraints

A basic assumption that is generally made about autonomy is that no one is always perfectly autonomous. No one is always capable of choosing a plan for himself or herself free from internal or external constraints.[14] We tend to recognize, however, that persons are capable of being substantially autonomous in their decision making. Yet certain internal constraints do serve to limit an individual's capacity for autonomous choice. The important question is: At what point do they make the elderly person incapable of making self-determined choices? The following case situation explores this question in the home care of an elderly couple who no longer seem to be able to live independent of nursing care and supervision.

Case #1: Aging Parents and Familial Responsibilities[15]

Cheryl Forbes is a Home Health Nurse who once took care of 82-year-old Mr. Fred Sims following his hospitalization for cardiovascular problems. Mr. Sims' condition improved under home care, and he was eventually discharged. Now, eight months later, Cheryl is frantically called by Mr. Sims' married daughter. It seems that Mr. Sims fell at home during the past week, but refused to see a physician. The daughter learned about the fall when her 81-year-old mother called her at work, and pleaded with her to come to the home and check on her father. Since this was the third such phone call from her parents during the past month, the daughter was beginning to wonder if she should consult with someone in making some long-term decisions for the care and supervision of her parents. She remembered Cheryl from her previous visits and hoped that she could help her with her parents.

Cheryl remembered that Mr. and Mrs. Sims lived alone in a small home. Their married daughter helps them buy groceries and accompanies them to their health care appointments. She also visits with them several times per week. Another daughter lives in a nearby town, but has chronic health problems that prohibit her active involvement in her parents' affairs. The Sims' only son lives on the West Coast and travels constantly in his line of business. He supports his parents by sending money for their expenses to his sister. All three children are apparently concerned about their parents, but only the married daughter is really available to assist them.

The present problem with the Sims seems to be caused by the fact that they are losing their abilities to live independently and make their own decisions. In the past, Mr. Sims was always the decision maker in the family. He strongly rejected any suggestion that he and his wife change their mode of living as they became older. He also rejected any formalized help from any of the children except to "make things easier for my wife." Now he is being asked to reconsider his decisions in light of his increasing unexplained falls, and the worries that his wife and married daughter are having. Both Mrs. Sims and her daughter hope that they can appeal to the nurse to help them make a decision that will preserve some autonomy for the couple. Yet it is doubtful that what is best for all concerned can avoid infringing on the choices and self-respect of Mr. Sims, in particular. Cheryl wonders if there is a happy medium for aging parents when it becomes obvious that they can no longer live independently. What is the role of the home health nurse in assisting individuals to reach these decisions?

Obviously, Mr. and Mrs. Sims are elderly individuals whose capacities for autonomous decision making are beginning to be compromised. The critical decision, both for the couple's children and the home health nurse, is whether they will treat the Sims as autonomous agents. If they do and if they are convinced that it is in the Sims' best interests to change their living arrangements, they may try to persuade them of the need for a change. They may argue for this change for appropriate reasons related to the Sims' daily care and safety.

If the children and the nurse have doubts about the Sims' capacities to be substantially autonomous decision makers, they may try to assess their parents' abilities more specifically. They may try to determine if the Sims comprehend the risks, the alternatives, and the advantages and disadvantages of various options. Again, their reasons for this approach are related to the Sims' daily care and safety, and the probability that some of their decision-making capacities might be impaired.

If one or both parents seem incapable of making reasonably autonomous choices, then the children and the nurse will need to decide whether or not to take over the decision making for the parents. If they do, several legal and ethical problems will occur. Legally, none of the children have any authority to take over decision making for the parents even though they might be well motivated. Certainly, the home health nurse has no such authority. The only agency with the legal authority to declare the Sims incompetent to make decisions for themselves is a court of proper jurisdiction. Hopefully, this move will not be necessary to plan appropriate care for the Sims and to help them to make decisions for their future care.

Depending on which approach the children and the nurse take with Mr. and Mrs. Sims, other questions also need to be answered. Is the married daughter (or any of the children) really in the best position to know the advantages/disadvantages of a more protected living arrangement for the Sims? Can one adequately choose for another, especially parents, where matters such as lifestyle and housing are concerned? The Sims are really in the best position to know the psychic trauma a major lifestyle change would cause them. How can someone else adequately decide for another where basic lifestyle and comfort are concerned?

It might also be prudent to ask whether or not deciding for the Sims would be a statement that the elderly couple have lost or are close to losing control of themselves. Mr. Sims, in particular, already seems distressed by that possibility. Deciding for a more protected environment might add considerably more stress to his daily living than currently exists in the familiar environment.

How does one really decide that the Sims' capacities for autonomous decision making are significantly compromised? Is there some point along the continuum of capacities for autonomous decision making where autonomy is recognizably compromised? Collopy suggests that autonomy be classified into two types: *decisional autonomy* is the ability and freedom to make choices; *executional autonomy* is the ability and freedom to carry out and implement choices.[16] Like many elderly individuals, the Sims have decisional autonomy, but not complete executional autonomy. The scope of their autonomy is, therefore, limited. Incomplete executional autonomy makes many elderly individuals vulnerable in their home environments because they are essentially dependent on others to execute their choices. The lack of executional autonomy also creates the situation where decisional autonomy is easily eroded once the elderly person's executional autonomy diminishes or is eventually lost. In such situations, self-determination (the underlying moral value of autonomy) is endangered. The potential for this type of situation developing in the life and care of Mr. and Mrs. Sims is a genuine concern.

Assessing the weight and priority of internal constraints on decision-making capacity in the life of elderly individuals living at home is very difficult. The Sims' situation indicates that it is not really clear when autonomy becomes so compromised that someone else may legally and ethically take over. However, the nurse visiting in the home can readily see when executional autonomy of elderly patients is being compromised and make needed interventions to prevent the erosion of decisional autonomy. Visiting nurses are an important health care resource for elders as their autonomy decreases because of internal constraints. They can help them and family members to make decisions that will protect elderly patients' decisional autonomy and provide assisted executional autonomy as long as reasonably possible in the home setting.

External Constraints

External constraints on elder autonomy can be manifested in several ways. Loss of funding for support services, such as transportation and community-sponsored programs, leaves some elders without opportunities to socialize or seek access to other resources in their communities. Legal requirements and institutional regulations may also limit the choices and opportunities of the elderly. The following case demonstrates the loss of autonomy experienced by one woman when state regulations required her to transfer to another nursing home.

CASE #2: WHEN PUBLIC ASSISTANCE LIMITS YOUR CHOICES[17]

Rosa Wilson is a 74-year-old woman who has been a resident of Southside Care Center for six months, following hospitalization for a CVA. She no longer needs skilled nursing care, but would like to live at Southside for the rest of her life. Mrs. Wilson has no children, and her only relative is an older sister who is confined to a nursing home in another state. She is mentally alert, but needs assistance with dressing, walking, and toileting. She receives public assistance for her care.

The present problem concerns Southside's classification as a skilled nursing care facility. Because Mrs. Wilson no longer needs this level of care, the state has requested that she be transferred to another long-term care facility. Southside's administrator does not like to move Mrs. Wilson, but must do so because her public assistance will no longer pay for the higher priced care that Southside offers to patients. Mrs. Wilson and four other publicly supported patients will be transferred to a facility that offers more traditional nursing home care.

After learning of the planned transfer, Mrs. Wilson calls a public assistance attorney, and asks him to represent her and the other patients in a legal suit to block their transfer. She argues that the state cannot move her and the other patients without a hearing about the benefits of their present level of care, and the potential harms and benefits that might occur at another nursing home with a lower level of care. Despite the fact that the state pays for all of Mrs. Wilson's care, she feels that she should have some input into the decision about her level of care and the selection of the new nursing home.

Like Mr. and Mrs. Sims, Mrs. Wilson has a physical disability that serves as an internal constraint on her autonomy. She also is subject to an external constraint in the form of state regulations concerning payment for her level of care and living arrangements. Since the policy of Southside Care is to provide skilled nursing care and Mrs. Wilson does not need this level of care, she is being transferred against her will to another institution that provides a lower level of nursing home care. Her autonomy is slightly compromised by her physical disability, but is greatly compromised by policies that do not allow her to choose the lifestyle she prefers based on her own internalized value system. As a result, decision making for Mrs. Wilson shifts away from a focus on what she would prefer and find comfortable to what is available within the constraints of state regulations. She has essentially lost her autonomy to choose the lifestyle that she finds desirable for herself.

It is unclear whether professionals who work in environments like Southside Care have an obligation to be advocates for patients like Mrs. Wilson and the others who will be transferred. If a basic responsibility of the nurse is to promote the health, safety, and welfare of the patient, then it seems that any nurse working in the institution should be concerned about the situation of Mrs. Wilson and the other patients. The transfer to another institution with a lower level of care might not be in Mrs. Wilson's best interests. She is comfortable with her present living arrangements and has improved her health status since arriving at Southside Care. She is able to take care of herself except for assistance with some of her daily living. Certainly, the move to a new and unfamiliar environment will present changes and differences that will take Mrs. Wilson time to adjust to and that may, in the long run, hinder her capacities for self-care at the present level. Since the number of nursing staff will be smaller in the new facility, there is reason to believe that Mrs. Wilson might decline in her mental and physical abilities when she cannot receive the level of care that she is accustomed to receiving. All of these possibilities seem to indicate that the nurse has a responsibility to promote Mrs. Wilson's welfare by resisting the transfer. At the very least, the nurse should ascertain the level of care that can be expected at the new facility and make sure that Mrs. Wilson will not likely suffer any decline in her new surroundings. Being an advocate for Mrs. Wilson means to protect her abilities to voice her own choices and to promote her welfare as the patient has defined it.

Overriding Elder Autonomy

Sometimes the obligation to respect elder autonomy can conflict with other obligations of the health professional. When decisions need to be reached about patient care, the nurse is obliged to make choices that will enhance the patient's health, safety, and welfare. Even in those situations where the nurse is not the decision maker, the nurse is obliged to provide information to the patient that will enhance the patient's choices.

This obligation to promote the best interests of the patient, however, often conflicts with the obligation to respect patient autonomy. If both the nurse and patient agree on what promotes the best interests of the patient, no conflict occurs. If the nurse believes that what is best for the patient and what the patient is choosing are not the same thing, then conflict is inevitable. In these situations, the nurse must then determine whether or not his or her obligation to promote the best interests of the patient is of greater moral weight than the obligation to respect self-determined choice. The following case situation illustrates this problem in the planning of care for an elderly woman returning home after hospitalization.

CASE #3: WHEN CHOICE OF LIFESTYLE
IS NOT IN THE PATIENT'S BEST INTERESTS[18]

Thelma Jenkins is a 72-year-old woman who has been in the hospital for three weeks. She was admitted for evaluation of a head injury following a fall in her apartment building. Her head injury has healed, and she is now receiving treatment for mild hypertension and mild orthostatic hypotension. This is the first time that she has received major medical treatment for any reason for over ten years.

In planning for Miss Jenkins' discharge home, her primary care nurse, Susan Betts, has arranged for a home evaluation. She learns that Miss Jenkins lives alone in a one-room apartment furnished with a bed, refrigerator, table, chair, lamp, and a small sink. Since she does not have a stove in her apartment, two meals per day are supplied by her landlord. She also eats her noon meal three days per week at the Senior Citizen Workshop in her neighborhood. With the support of her social security check and food stamps, she has always had adequate money for her needs and has lived for over ten years in this apartment.

Ms. Betts needs to make sure that this elderly patient can adequately maintain herself at home, take her medications properly, and maintain a low-salt diet in her home. Since Miss Jenkins has no relatives and essentially cares for herself, her mealtime arrangements are a concern. Ms. Betts calls her landlord to see if she is willing to come to the hospital and receive information about Miss Jenkins' diet. The landlord is not very friendly and seems hesitant to agree to prepare a low-salt diet for Miss Jenkins. She also complains to Ms. Betts that Miss Jenkins' apartment is not very clean, a fact that concerns not only the landlord, but Miss Jenkins' neighbors. Apparently the long-term accumulation of dust and dirt in the apartment has attracted cockroaches, which are bothering her neighbors. The neighbors have also reported that Miss Jenkins keeps an open bucket of toilet wastes in her apartment (she shares a hallway bathroom with two other tenants in the building). Although she empties this daily, the neighbors feel that this also attracts pests to the apartment. The landlord acknowledges that Miss Jenkins is a regular paying tenant, but she seems disappointed that she is returning home.

When Ms. Betts reviewed Miss Jenkins' admission note from the ER nurse, she noticed that the patient had been described as physically dirty with unwashed hair, ragged and dirty fingernails, and overgrown toenails. The clothes she was wearing on admission were also described as dirty and unkempt. When Ms. Betts questioned Miss Jenkins about her lifestyle and living arrangements, the patient stated that she was very comfortable in her apartment, that she liked her neighborhood, and that she felt safe there.

Ms. Betts suggests that Miss Jenkins might find it desirable to find different living arrangements, but the patient soundly rejects this idea. She is comfortable with her lifestyle and simply does not want to make any changes. The nurse next offers to contact agencies—homemaker service, senior citizen's groups—to help Miss Jenkins clean and maintain her apartment. Again, Miss Jenkins rejects these ideas. She says that she is comfortable and that she does not want (or need) help from anyone. She is very eager to return home.

Ms. Betts is convinced that Miss Jenkins can probably maintain her health and continue her medication and low-salt diet routine better in a more protected environment. She would even go so far as to say that it is in Miss Jenkins' best interests to do so. Yet she is uncomfortable with using her role as the discharge planner to set in motion interventions that might eventually result in the patient's loss of the right to control her person, her financial resources, and her environment. Can an individual in the community be forced to be cleaner and to live in a clean environment? Should these be conditions for discharge from the hospital? How far does the nurse go in promoting "good" for patients, and who determines what is "good" for Miss Jenkins?

The nurse caring for Miss Jenkins has a difficult decision to make. Would Miss Jenkins be better off in another living arrangement? She might be better off medically where her hypertension is concerned, but she might not be better off where her subjective choices and values are concerned. The patient seems to prefer her familiar apartment and the amount of control that she can exercise within it. This is ultimately important to her sense of well-being. This sense of well-being is not dependent on the cleanliness of the apartment or the absence of cockroaches.

Some might argue that Miss Jenkins would be better off living in a cleaner environment, and having her clothes and grooming taken care of. Should the nurse be concerned about this? Is the nurse obliged to take the *total* welfare of the patient into consideration, including such personal items as house cleaning, grooming, and maintenance of suitable clothing, or is the nurse merely obliged to consider the patient's welfare in a narrower sense, say, for example, her diet and blood pressure readings and her *health* welfare?

Most would agree that no health professional can ethically or legally override a patient's choices in order to provide benefits in the form of a clean environment. Yet many elderly citizens are victims of this type of intervention as long as it can be argued and supported by law. If it could be argued that Miss Jenkins' apartment is a health hazard because of its uncleanliness and cockroaches, then she might be prevented from returning to live there. If it could be argued that most people would not want to live in this type of environment, Miss Jenkins' desire to do so might become suspect. Her ability to make these types of choices for herself would be in question, and her future choices would be in jeopardy.

The value of well-being, as defined by the patient, is an important consideration in situations involving the autonomous choices of elderly individuals. Although promoting well-being is an obligation of the nurse, in most cases, well-being must be defined by the individual and not the nurse or another health professional. Yet the conflict of well-being, as defined by someone on the health care team, with the autonomy of elderly individuals is one of the most frequently mentioned ethical conflicts by those who care for the elderly.[19] Resolving these conflicts is often a difficult ethical matter in health care decision making for the elderly individual.

Issues for Family Members

Health professionals and family members usually work together to protect, promote, and maintain elder autonomy in home-based, long-term, and acute care. Family members become important sources of information about the elder's values and previous choices, and often confirm values expressed by the elderly individual in conversations with the nurse.

Some families, unfortunately, know little about their elderly relatives' values and potential choices under specific conditions. They have never asked or no one has ever encouraged the elderly individuals to discuss such matters with family members prior to a health crisis. Every elderly individual can be assured that health professionals want to follow his or her choices and desires, to the greatest extent possible, if they know what they are. Hence, the elderly need to be encouraged to discuss their values and choices with family members, and to designate some family member or a good friend to be their surrogate decision maker as they become older.

Family members, on the other hand, need to be reassured that they are playing a very important role in the future health care of their elderly relatives when they become involved with their choices and values. To the extent possible, children ought to share responsibilities for their parents so that the burdens of taking care of elderly parents do not fall on the shoulders of one sibling more than the others. When children are not immediately accessible to their parents, other arrangements may need to be made as their parents grow older and have a greater need for assisted living conditions. The majority of elderly individuals can live well and in comfort if plans are made for daily living arrangements, and acute and long-term health care in advance. Once a health crisis develops, however, decision making may not be optimal because of unclarity about the elder's choices, and the costs of providing nursing home care or assisted living arrangements under the crisis situation.

When elderly citizens have no living relatives and are essentially on their own, health professionals have an added obligation to assist individuals in their planning, and to document their choices and plans for the future. Social workers and patient advocates often prove indispensable to elders entering the health care system by helping them to remain in their own living arrangements or to select another arrangement that is consistent with their values and financial resources. The nurse in the community provides a valuable service to social workers and discharge planners when they evaluate the home environment for its safety and comfort following a hospitalization, especially when the elderly citizen temporarily loses mobility and independent living ability. Although it is always desirable for the elderly to remain in their own homes and communities, some will not be able to do so because of compromised physical and mental abilities and associated loss of autonomy. When this occurs, care givers, both health professionals and family members, need to consider how the overall well-being of the elderly individual can be promoted within the family setting.

Some well-being can be promoted by maintaining whatever autonomous decision making the elderly individual is capable of. Overall well-being is best promoted by acting in the best interests of the elderly individual and, in some instances, by overriding elder autonomy when the expression of individual choice is clearly not in the individual's best interests. Acting in the best interests of elders may be very difficult for family members who have previously fought to keep their loved ones autonomous. Again, nurses and other health professionals often play a vital role in assuring family members that they are acting responsibly and morally on behalf of their relatives. To the extent that these mechanisms protect, promote, and maintain the executional and decisional autonomy of our elderly citizens in health care decision making, there is reason to believe that such decision making will be ethical in nature.

References

[1]V. R. Fuchs (1984) Though much is taken: Reflecting on aging, health, and medical care. *Milbank Memorial Quarterly* **62(2),** 160.

[2]D. Callahan (1987) Restructuring the ends of aging, in *Setting Limits: Medical Goals in an Aging Society.* Simon and Schuster, New York, NY, p. 21.

[3]N. Froner (1985) Old and frail and everywhere unequal. *Hastings Center Report* **15(2),** 27–31.

[4]D. Callahan, What do children owe elderly parents? *Hastings Center Report* **15(2),** 32–37.

[5]W. F. May, Who cares for the elderly? *Hastings Center Report* **12(6),** 31–37.

[6]Froner, *Hastings Center Report,* 29.

[7]D. G. Blazer (1986) Suicide in late life: Review and commentary. *Journal of the American Geriatric Society* **34(7),** 519–525.

[8]M. Tolchin (1989) When long life is too much: Suicide rise among the elderly. *New York Times,* July 19, pp. A1,A15.

[9]V. R. Fuchs (1983) *How We Live: An Economic Perspective on Americans from Birth to Death,* Harvard University Press, Cambridge, MA.

[10]T. L. Beauchamp and J. F. Childress (1989) *Principles of Biomedical Ethics,* Oxford University Press, New York, NY, pp. 71–74.

[11]G. Dworkin (1988) *The Theory and Practice of Autonomy,* Cambridge University Press, New York, p. 10.

[12]Dworkin, *The Theory and Practice of Autonomy,* p. 20.

[13]R. M. Veatch and S. T. Fry (1987) *Case Studies in Nursing Ethics,* J. B. Lippincott Company, Philadelphia, PA, pp. 101–105.

[14]Veatch and Fry, *Case Studies,* p. 102.

[15]Veatch and Fry, *Case Studies,* p. 102 (case adapted with permission).

[16]B. J. Collopy (1988) Autonomy in long term care: Some crucial distinctions. *The Gerontologist* **28(suppl.),** 10–17.

[17]Veatch and Fry, *Case Studies,* p. 91 (case adapted with permission).

[18]Veatch and Fry, *Case Studies,* p. 59 (case adapted with permission).

[19]R. A. Kane and A. L. Caplan, eds. (1990) *Everyday Ethics: Resolving Dilemmas in Nursing Home Life.* Springer Publishing Company, New York, NY, pp. 48,49.

Rethinking Family Loyalties

Evelyn M. Barker

Family Loyalties and the Common Good

Family loyalties play an important part in decision making on life-extending therapy: When told a child's life is at stake, parents feel obliged to agree to even the most experimental, high-risk measures. A son may demand life-extending measures for an elderly parent he would not want for himself. A new widow comforts herself with the thought that everything medically possible was done to save her elderly husband. Especially in the early and late stages of life, the seriously ill depend on the devotion of family to safeguard their interests in life and health.[1]

Recent proposals to exclude elderly from high-technology life-extending therapy challenge these traditional loyalties. They argue from an abstract rational viewpoint focused on the common good, scanting or bypassing family ties. Daniels argues that a prudent person would prefer to maximize an individual's opportunities to reach a normal lifespan, rather than to extend it beyond a normal human life course, and will consider just a distribution of society's limited health care resources that reserves expensive high-technology care for younger age groups.[2] Callahan maintains that a philosophical understanding of the meaning of old age implies that disengagement from social life be accompanied by acceptance of death on the completion of a normal life course.[3]

Neither has generated a rational consensus on the goal of medical care for the elderly: The practical implementation of a social policy to exclude the elderly from expensive high-technology medicine is thwarted not only by the political power of the elderly, but by the strength of family loyalties to older age groups. A goal of geriatric medical care that violates norms of family loyalty is unlikely to be morally acceptable or politically viable. To meet democratic standards, both older patient and kin must relinquish claim to life-extending treatment.

In the past, medically possible measures were limited, and the family bore the expense of treatment. Here, the maxim to "do everything possible" embodied a morally admirable expression of family loyalty, reinforcing the medical ideal of prolonging life. Today, however, artificial life supports can sustain those in a persistent vegetative state for years, whereas antibiotics permit others with minimal functioning to linger indefinitely within institutions. Moreover, the cost associated with such care is borne by us all. On these grounds, the ethical maxim needs questioning, just as the medical ideal needs qualifying. The life-extending options of contemporary medicine demand a rethinking of family loyalties in order to confront death realistically. In this era of technological medicine, a more appropriate maxim is: "Do what is worthwhile."

In examining the content of family loyalties, I mean to stress the positive role they deserve in setting goals for medicine. The family is the social institution most capable of dealing with serious illness and death in a way that lessens their evils, and may even redeem them. The ethos of mutual dependence within family life makes a haven for one whose bodily condition deprives one of self-reliance. The rational individuals of contemporary ethical and social theory are robust adults, but medicine must take account of the role of the family in coping with human weakness and vulnerability.

Plato inaugurated philosophic criticism of the family as an "exclusive center of joys and sorrows" that impedes the rational reorganization of society needed to establish justice and promote human progress. According to contemporary contract theory, the partiality of family members stands in the way of a fair or egalitarian distribution of social resources, and needs to be countered by the viewpoint of a detached rational ego.[4] According to this approach, independent individuals make a social compact with other solitary individuals, whereas the family is a ghost hovering over individuals' plans without a distinct role in the social organization.

Josiah Royce, seeking to develop a distinctively American ethic embodying democratic principles, makes a different evaluation of the family. He found in family loyalty the source of a morally enlightened individualism essential to a democratic society because it generates other-regarding selves rather than self-absorbed egos. To unite a culturally diverse population into a coherent democratic community, he proposed an American ethic based on family loyalties.[5] This chapter invokes Royce to challenge much touted individualist ethics and offer a model for decision making that incorporates family loyalties.

An Ethic of Family Loyalties

Family loyalties link one with others who play a morally relevant role in one's own life. They usually operate forcefully in ethical decision making, powered by affection for an immediately perceived individual closely identified with one's own past. Confirmed by religion and social tradition, the family bond is a deep source of moral obligation, the violation of which causes shame, guilt, and bad conscience. In this loyalty, awareness of a biological bond is strengthened by shared experiences within family life. These combine to make the good of one family member an integral part of the good of another. Consequently, the ill-being of a family member disturbs the equanimity of others within the family circle.

In a crisis medical situation, kin loyalty can be excited in ways not conducive to reasonable decision making. Family members may focus solely on obtaining medical measures to forestall death. Heidegger characterizes the attitude of both patient and family toward death as evasion.[6] Thus, patient and family regard death as an accident that happens to others, and to oneself in an indefinite future—not now. To think of death, particularly of one's own death, is morbid, and to be avoided. When death obviously looms for another, the discomfort of intimates leads to the isolation of the mortally ill.

Heidegger attributes this attitude to the way we image the world of daily life: The world is a workshop in which we perform tasks—a playing field in which we compete with others for goods and prizes.[7] It has no place for those out of action. Here, we reduce the seriously ill person to a malfunctioning body—an object for medical technology—and the patient views himself or herself as running a race to outstrip death, marshalling all forces at his or her disposal.

Such fear of death precludes deliberation and prejudges the issue. Consequently, a changed attitude toward death is the key to enlightened decision making in medical care of the elderly. Heidegger's existential phenomenology applies particularly to old age, when death becomes a rational expectation rather than a remote contingency. Since Heidegger makes anticipation of death the key to authenticity in life, his views provide a base for an ethic of family loyalty applicable to the elderly.

Heidegger insists that a true understanding of human life starts with the recognition that death cannot be outstripped.[8] Victory and peace come by anticipating death rather than letting it overtake one. Anticipation of death is a positive act, for only then can persons see their lives as a whole and interpret genuinely their present situations. Envisaging life as a whole requires thinking past one's own death to the meaning of one's whole life to both oneself and others. Just as one's own life is a part of the life of others, so the lives of others are part of one's own.

From this vantage point, the future of an older person is a limited period, bounded by death, in which the older person needs to attend to what is still outstanding in life, one's personal unfinished business. This includes what the individual can be for others who are a part of one's own life, and what others can be for oneself. By confronting the inevitability of one's own death and that of others, one stands in a better relationship to them. Just as one must die one's own death, so one ought to live one's own life, neither exploiting others nor letting another take over one's own life. The fact that the future holds death for each means that each must make room for the other in the present.

This vision fosters the other-regarding individualism Royce stresses. Enforced separation from loved ones in death demands the preparation of individuality and independence within daily life. Separateness does not imply indifference. Instead, it involves a mutual recognition of a limit to what each can be for another, a limit set by nature in human mortality.

In medical settings, family members express concern for one another by seeking goods and services members can use in their present situation. When a member is in difficulty, one tends to step into his or her place to solve the problem for him or her. Heidegger calls such solicitude defective, for one makes the other dependent, and both lose their freedom. In a freedom-gaining relationship, one looks ahead with the other to help him or her understand what lies in store and cope successfully on his or her own. One steps back to let the other deploy his or her powers and steps forward to support in times of weakness, leaving the other free to determine his or her own fate.[9]

In old age, one's future always has death in store. Although no one can step into one's place here, one can help another to achieve readiness for death. Readiness for death is not achieved by resignation from life or willingness to die, but by attending to what remains outstanding in life. This encompasses not only what remains of one's own future life, but one's part within the lives of intimates.

Although retirement may disengage one from the work force, old age does not dispense one from participation in the ultimate purposes of social organization—the meaningful lives of individual human beings. On the contrary, old age liberates the individual to concentrate on these ultimate purposes, rather than the short-term outcomes that tend to dominate our working mentalities. The meaning of old age is not to be sought in a new social role as conservator of the human past (as Callahan suggests), but in freedom to attend to "what is still outstanding" in one's own life as a privilege of a lifetime of service.

The social role the elderly should assume is not to step aside for younger generations, but to lead the way in being mindful of the intrinsic goods of human relationships. Heidegger ascribes to "authentic consciousness" a willingness to be "the conscience of others."[10] This notion may grate on our liberal mindset, suggesting the forcing of one's values on others. Yet for Heidegger, it does not imply this, but implies, instead, encouraging others to be mindful of the whole of their lives, by being ourselves attentive to the whole life of each. In this orientation to another, we assess our common past as something to be remedied for its shortcomings, preserved (and even enhanced) in value in the future we enjoy together.

Family Loyalties

Readiness for Death

"Empowering" the elderly to make their own medical decisions conforms with an ethic of family loyalty. The reluctance of family members to empower elderly relatives, and of the elderly themselves to assume decision-making roles, may reflect habits of negative dependence.[11] However, dependence also may be positive and appropriate when old age impairs one's capacity to be a judicious consumer of medical services. Gradually declining powers may make it difficult for the older person to tell whether he or she has benefited by therapy. Furthermore, a life-threatening illness usually places an older person in a new kind of dependence: instead of mutual dependence, one-sided dependence of indefinite duration, perhaps only ended by death rather than recovery. The probability of such dependence on one's intimates means that the future course of others' lives is implicated in one's medical decision. It is neither irrational nor a sign of moral weakness to want intimates to participate in decision making, even to make a choice for one.

Family members rightly play an important role in decisions concerning life-extending therapy for an older person, since its success requires their cooperation and its undertaking may have a major impact on their lives. The issue is whether family participation is authentic and freedom-gaining or inauthentic and freedom-losing. In the freedom-gaining mode, family members help an older patient (1) achieve readiness for death and (2) choose medically and morally responsible options regarding life-extending therapy.

Out of guilt for previous neglect or fear of appearing remiss, relatives may press hard for life-extending measures as a way of making amends or proving loyalty. They may collude with the patient in refusing to acknowledge the imminence of death, or hold out its possibility as a reason for a reluctant older person to undertake arduous and futile medical measures. Such behavior clearly is freedom-losing, taking from both patient and family member the freedom to choose authentically.

Another trope is to concentrate on the older person's immediate physical needs, laying on the services of others while ignoring the patient's state of mind. Attending to a patient's immediate physical needs is important, especially when it assures the comforting presence of an intimate, but the unique and irreplaceable role a family member can assume in a crisis medical situation is to help the patient achieve freedom for life and death.

Understandably, family members are chary of alluding to the possibility of death. An older person may fall into a deep melancholy and become withdrawn, facing a future of pain, physical indignity, and the loss of valued activities. Relatives must deal with these fears, and with their own pain and anxieties in response to them, but since death is *always* in the near future of older persons whether undertaking further medical treatment or not, it needs to be prepared well before medical crises.

In many respects, the current ethos of medical care supports freedom-gaining relationships between older patient and family. The new emphasis on truth telling leads doctors to be candid (even brutal) in informing terminal patients of their conditions and prognoses. In order to satisfy principles of informed consent, health professionals do not conceal the probability of death as an outcome of medical intervention. Recent criticisms of medical paternalism inhibit a doctor from choosing among alternative courses of action for the patient. Finally, hospitals now discharge patients as soon as possible for financial reasons, forcing both patients and families to confront how patients are to be cared for at the end of life.

Current legal devices of Living Will and Medical Proxy can also open up the subject of the medical management of the final stage of life among family members. In the Living Will, individuals give advance directives for withholding or stopping available life-sustaining procedures. In Medical Proxy, individuals designate another to make such decisions on one's behalf. Although a majority of states have legalized such documents, they are not widely used nor are their powers well understood.

These medical and legal approaches can serve to transform the question of therapy for the elderly in a realistic way more amenable to shared family decision making. Hitherto, the issue has been framed as an exclusive life/death alternative, therapy standing for the "life" option, and no therapy for the choice of death. With the Living Will and the Medical Proxy, the question becomes, instead, one of permitting or rejecting specific medical measures when one's condition is known to be terminal. In reality, a decision about "life-saving measures" is a decision about whether to live on in expectation of death either without treatment or after treatment. Death may occur sooner because of medical intervention, and later without it. Thus, family members need to help an older patient cope with a future that always has death in store. Family members need to be responsive to the patient's overtures and tactful in leading another to be free *for* death.

Living Wills are sometimes called permits to "pull the plug," but they are also able to assure desired treatment, and to discriminate wanted from unwanted medical measures. For example, one may opt to continue kidney dialysis but reject resuscitation after cardiac arrest. Since life-extending medical technology is generally available only within hospitals, staff can readily institutionalize the Living Will in medical practice, much as consent forms are now routinely obtained. While relieving health professionals of fear of negligence suits, such steps also prepare a family for a crisis.

Making medical arrangements positions both older patients and intimates for a personal orientation to the end of life. The patient who recognizes the inevitability of death does not just lay his or her head on a pillow and turn his or her face to the wall. One looks to settling accounts with those who matter or to whom one matters. Expectation of death provokes attempts to heal the imperfections of the past by reconciliation of differences with others, revelation of what ought to be told, righting of old injuries, or repentance. It inspires the sealing of life's perfections by expressions of gratitude and recognition.

Through concern for the future of others after one's death, one secures a place in their ongoing lives. Although death puts an end to one's own possibilities, it happens within the lives of others, affecting their ways of meeting life and death. The final gift of one to another may be a "happy death," showing wisdom and courage in dying well.

Choosing Life-Extending Therapy

The champion of life-extending therapy for the elderly should not underestimate the rigors of high-technology treatment. Already, the use of life-support systems has produced right-to-die legislation to avoid respirators and eliminate artificial feeding through legal documents, such as the Living Will. Even when medically successful, a high incidence of mental depression, suicide, psychological problems, divorce, and family conflict may shadow the extended lives of patients and their families. For example, in arduous therapies, the patients may be required to pursue a regime that controls the life pattern for the whole household indefinitely. When those over 65 undertake such therapies, they will need the daily companionship, encouragement, and ministerings of intimates for decades. Two senior generations may rival one another as well as a younger generation, not only for family funds, but for the time and attention of family members.

Before we, as a society, endorse high-technology life-extending therapy for the elderly, we need to think about its effect on the family, as well as the value of an extended life plan for the elderly. Will the old be exploited by others or exploit family members in their pursuit of an extended lifespan? Traditionally, the onset of old age has justified the lessening of one's obligations, permitting the old leisure for personally meaningful activities. Can we achieve wisdom and serenity in unending technological combat with the aging process? Will we elderly show generosity and consideration for the future of others, or each be fixed on her own fate?

An extended life plan will preserve family loyalties only if it permits freedom-gaining relationships, rather than freedom-losing ones to characterize family life. These reservations do not negate high-technology therapy for an older person in all circumstances, but they do point to conditions other than medical indications that involve family members in decisions.

A strong-minded affluent elder may cope independently with minimum reliance on family, and command resources to pay for needed service. More often, successful therapy will require the devotion of a family member to supervise the older person's daily physical and psychical wants. A spouse to whom the loss of a life's companion would be a grievous blow is very likely to take on this role without even a sense of personal sacrifice. The personal affection of a well-placed adult child, grandchild, or sibling may also induce him or her to carry out this function, especially with support and regular respite from other family members or paid helpers.

An intimate involved in the older patient's aftercare may legitimately expect to take part in decision making about such treatment. Far from interfering with the patient's autonomy, counsel with family members is essential to wise deliberation, even when the patient is quite competent. In medical situations, there is a tendency to think that even a low probability of a favorable outcome is a worthwhile choice if death is the alternative. What this ignores is the relevance of the high probability of very unfavorable outcomes to those other than the patient.

Interested relatives are best placed to take account of the fact that the distress of therapy and recovery must figure as capital outlay in the case of the older person, and be measured as a demerit in the calculation of benefits, not passed over in silence. Although doctors are likely to note the death risk during surgery itself, they pay less heed to death risks and discomforts from complications or side effects of therapy.

What counts as a high probability of a favorable outcome? As in other decisions, not 1 out of 10, or even 50-50, but an outcome that is distinctly more probable than improbable. Relatives need to focus on unfavorable possibilities, because there are many pressures on medical professionals not to dwell on them. In addition to a disposition to take a cheerful view of a patient's prospects for the patient's own sake, doctors and hospitals are motivated to recommend treatment rather than not. The whole medical-industrial complex is geared toward medical intervention in the case of elderly patients by the Medicare program as well as its own dynamics, so that it is left to the family to question how likely therapy is to promote the well-being of the older patient during his or her remaining life course.

As major medical interventions become acceptable in older patients, there is a risk that older patients will become subject to abuse in experimental therapy. Some older persons go along with therapy offered out of fear of being abandoned by their doctors should they refuse. Once again, relatives can protect an older patient from unwittingly or delusively participating in experimental therapy, doing so without full understanding of the risks involved, or not exercising the right to withdraw from experimental therapy or refuse it when offered.

Lastly, relatives are well placed to determine whether an older patient's personal situation favors life-extending therapy. Intimates can tell whether he or she has the ego-strength to endure its vicissitudes with equanimity. Adult relatives who will sustain the patient during a protracted recovery must assess their resources. A family member must be candid about the degree of responsibility he or she can take for the patient's care during the time of greatest dependence. This information is a vital factor, positive or negative, in the comparison of benefits and drawbacks of alternate therapies or no therapy.

Doing "everything possible" will not better the whole course of an older person's life if it divides family loyalties by placing an unacceptable burden of unlimited duration on an adult relative, especially one with other family responsibilities. What is acceptable will depend not only on the demands of the patient's condition, but on the health and character of the intimate and the role each plays in the life of the other. Doing "what is worthwhile" acknowledges that death sets a natural limit on what each person can and ought to be for another, whether parent and child or husband and wife. This maxim brings family loyalty into harmony with current medical ideals on the prolongation of life and is in accord with democratic principles respecting the dignity of each human life.

Notes and References

[1]This chapter expands a talk presented at a University of Maryland Baltimore County conference, Oct. 1988, sponsored by the Maryland Humanities Council.

[2]N. Daniels (1988) *Am I My Parents' Keeper*. Oxford University Press, New York, NY.

[3]D. Callahan (1987) *Setting Limits: Medical Goals in an Aging Society*. Simon and Schuster, New York, NY.

[4]J. Rawls (1971) *A Theory of Justice*. Harvard University Press, Cambridge, MA, is a well-known example.

[5]J. Royce (1914) *The Philosophy of Loyalty*. MacMillan, New York, NY.

[6]M. Heidegger (1962) *Being and Time*. Harper and Row, New York, NY, sections 51 and 52.

[7]Heidegger, *Being and Time,* sections 15, 18, and 27.

[8]Heidegger, *Being and Time,* section 53.

[9]Heidegger, *Being and Time,* section 26.

[10]Heidegger, *Being and Time,* section 60.

[11]*See* M. B. Kapp (1989) Medical empowerment of the elderly. *Hastings Center Report* **19(4),** 5–7 for discussion of these problems.

The Role of Intimate Others in Medical Decision Making

Nancy S. Jecker

Individual autonomy has been a fulcrum of bioethical debate since the 1950s and 1960s, and a guiding idea behind a diverse body of bioethics literature. Dominant ideas, such as individual autonomy, pose the risk of creating conventional categories of thought to which society becomes wedded. Such categories may ignore central aspects of moral experience, thereby fostering illusions that become difficult to dispel. Intimate associations are one domain of moral experience that may elude the world of value delimited by a traditional autonomy model.

Recently, several authors have called attention to problems that arise from neglecting this domain in law and ethics. For example, Rhoden doubts the wisdom of present legal approaches to termination of treatment for incompetent patients, believing they place undue burden on patients' families.[1] Referring to the same legal standards, Burt also detects inattention to the legitimate role of intimate others in medical decisions. Discussing *Quinlan,* a now famous New Jersey case, Burt maintains that the father of Karen Quinlan "could not view the treatment decision through *her* eyes, as the court directed, without deciding whether Karen would want to view her decision through *his* eyes. ...He could not...sustain the clear-cut distinction between self and other that the court enjoined."[2]

Blum[3,4] and Ruddick[5] have noted a similarly troubling pattern in post-Enlightenment ethical theories. The term *autonomy* entered moral philosophy in the eighteenth century,[6] when Kant interpreted it to mean individuals' abilities to be governed by moral laws that they themselves author. Inspired in part by feminist thinkers,[7,8] Blum, Ruddick, and other philosophers[9–12] now object to what they see as a growing tendency for contemporary theories to be dominated by abstract principles that view autonomous individuals as separate from all of their essential moral relationships. In particular, these critics charge that current theories downplay the significance of special moral relationships, such as friends and family.

Building on this work, this analysis focuses on intimate relations, examining the connection between intimacy and autonomy, with an eye to orchestrating these diverse elements. Specifically, this study shows

1. How the present emphasis on patient autonomy overlooks ways in which intimate relations enable autonomy to function meaningfully;
2. How traditional categories of *competent and incompetent* discount intimacy as a tool for accessing patients' subjective experiences; and
3. How intimate associations mark moral boundaries for autonomy because they constitute a setting in which persons give and expend finite human resources.

The primary focus is on the role of adult children in decisions affecting elderly parents. Although the implications of this discussion are noteworthy for other forms of intimate associations with patients, they are beyond the scope of this project.

Attending to Context

A distinguishing feature of doctor–geriatric patient relations is that often an adult child or spouse shepherds the geriatric patient to hospital visits. Adelman, Greene, and Charon note that relatives accompany geriatric patients to outpatient visits in 20% of cases at a major urban teaching hospital.[13] Whereas other patient groups, such as pediatrics and obstetrics, may be joined by third parties, when a third party attends an internal medicine consultation, the patient is generally elderly. Studies also document that adult children frequently are actively involved in home health care, and are otherwise assisting and caring for elderly parents.[14,15]

Likewise, family members frequently affect inpatient care for elderly patients in significant ways. For example, I recently observed a patient management conference discussing 14 patients on a geriatric ward of a major university hospital; for 13 patients, family members were mentioned. Their role was most frequently discussed in regard to patient discharge (10 cases), and often had a direct or indirect impact on medical treatment decisions (five cases). Less frequently, the role of the family came up in discussing how to resolve in-hospital patient management problems, such as feeding (two cases), general patient morale problems (one case), and financial difficulties (one case).

The lives of family members themselves are affected profoundly by elderly ill relatives. As Livingston observes, caring for a sick family member can be a career in itself; illness can mean an added financial strain and may force a change in housing or lifestyle, and in diseases carrying a stigma, the stigma often attaches to family members as well as patients. Moreover, relatives may react to a patient's illness by developing their own psychopathologies. For example, in stroke, a condition more common among older age groups, 62% of spouses have a high likelihood of significant disturbance in their adjustment to stressful circumstances, ability to cope, general physiological condition, moods, and social functioning.[16] Relatives of patients with myocardial infarction also evidence functional disabilities.[17] Also, on learning of a family member's diagnosis of a cancer that threatens survival, one individual reported, "It seemed easier to face death for oneself than to watch someone else..."[18]

Although a number of studies attest to the fact that adult children perceive themselves to be obligated to meet the needs of aging parents,[19,20] this is not always the case. Children may lack the interest or skill to be active players in parents' health care. Overwhelmed by its demands, they may reject filial responsibility altogether;[21] they may form coalitions with health professionals that ignore or discount aging parents;[22,23] they may push for overtreatment of relatives out of guilt or insist on undertreatment to escape onerous financial or emotional burdens. In many cases, however, a presumption that family members are good-intentioned and competent is appropriate. In such cases, interpreting their preferences as emotionally maladjusted may be patronizing or cruel.[24]

Given the pivotal role family members often play in medical care of geriatric patients, we would do well to review the dynamics of their filial relationships, especially the parent–child relationship. In the past, parents did not live long enough to see offspring enter old age. Today, however, those over the age of 75 are the fastest growing age group in the country, and they have adult children in the 55- to 70-year-old age range. Thus, the dramatic gains in average lifespan in this century have not only brought about an aging society, but also have transformed relationships between parents and children in later life. For the first time, these relationships involve elderly persons at both ends.

Relations between old-old parents and their young-old progeny are noteworthy in several additional respects. First, those in the birth cohort 75 and over represent the last survivors of massive waves of immigration that took place before the First World War; also, they represent the lesser wave of immigrant refugees from Nazi Germany during the 1930s. Roscowe describes this group as deeply ensconced in ethnic communities and institutions.[25] The very fact of their ethnicity functioned as a barrier to educational opportunities, which were limited already during the first three decades of this century. By contrast, immigrants' offspring grew up in a more pluralistic environment, an environment in which ethnicity was of diminishing importance in individuals' lives. Ethnicity was also losing a great deal of its structural significance in social organization relative to such factors as race, socioeconomic class, and gender. Furthermore, immigrants' offspring benefited from the tremendous boon in educational opportunities following the Second World War, most notably, the GI bill, which made higher education available on a massive scale.

The implications of these remarks for the ability of each group to deal successfully with the health care system are noteworthy. An adult son or daughter often will be in a position to bridge communication gaps between parents and health professionals, facilitate exchange of information, convey explanations parents will understand, negotiate treatment decisions, and offer emotional support in dealing with medical problems, treatments, and consequences.[26] Therefore, incorporating offspring into the decision process often will be justified on the grounds that doing so enhances the patient's own decisional abilities.

Even a well-educated, articulate, and fearless patient stands to benefit from a family facilitator. As Brody notes, the family is the forum in which the process of "trying on" and "bouncing off" ideas typically occurs: a patient "cannot know what values she holds until she goes through a process of 'trying on' various value stances and 'bouncing them off' others whose opinions and reactions she cares about. ...Values...emerge from the dialogue process..."[27]

Rhoden (1988) agrees:

> The family is the context within which a person first develops her powers of autonomous choice, and the values she brings to these choices spring from, and are intertwined with, the family's values. A parent may understand the child's values because she helped to form them, a child may grasp a parent's values because the parent imparted them to her, and a couple may have developed and refined their views in tandem.[29]

Gadow refers to this kind of facilitation and dialogue as "existential advocacy."[30] She describes it as the effort to help persons become clear about what they want to do by helping them discern their values in a situation and, on the basis of self-examination, reach decisions. Such decisions express a reaffirmed, perhaps recreated, complex of values. Only through dialogue, when persons are engaged and their values expressed and responded to, can decisions possibly be *self*-determined.

These thoughts ring true. Surely, shaping moral goals is no easy task. It is not a task many can execute confidently or successfully alone.

Intimacy and Access

Let us next consider the role of adult children whose parents are deemed incompetent to make autonomous health care decisions and have provided neither a clear indication of their wishes, such as a Living Will, nor a clear delegation of decision-making authority, such as a durable power of attorney.

According to the traditional framework concerning surrogate decision making, individuals are judged incompetent if they lack the capacity to give informed consent to medical treatment. In such cases, someone other than the patient chooses the best course of action. Surrogates employ different standards for making this determination. According to a standard of substituted judgment, surrogates strive to speak for the patient by making the choice that the patient would make under the circumstances. Alternatively, where the patient's preferences cannot be ascertained, a choice is based on a best interests standard. Unlike the substituted judgment standard, a best interests model does not depend on what the patient would want, but on an allegedly objective weighing of the burdens and benefits associated with various treatment options.

The framework for considering these issues already casts them in terms of bipolar categories. On the one hand, the patient is either competent or not: either the patient possesses a capacity to form autonomous choices or the patient's authority is transferred to others who do. On the other hand, either others can uncover evidence of the patient's wishes, or else efforts to gather evidence are abandoned and judgment is grounded in objective standards.

It is the very clarity of this framework that renders it problematic. By ignoring a significant gray area, which might be termed "marginal competence,"[31] this bipolar framework discounts the realities of both patients' and families' experiences. In this uncertain space, patients may be formally judged incompetent, yet able to collaborate with intimate others to construct choices consonant with their experience. Likewise, intimate family members may be unable to furnish objective proof of patients' wishes, yet may have access to a rich and detailed knowledge base from which they can build a sense of what constitutes harm or benefit to the patient.

One example of this is a daughter who serves as caregiver for a frail elderly parent. She may be able to look beneath the disoriented conduct and broken phrases to construe or help create some semblance of what the parent values. Evidence is gleaned through partaking in daily rituals with the parent, such as bathing and feeding, and interpreting the parent's responses. Just as a parent comes to know an infant by observing the infant's posture, sounds, eye movements, and facial grimaces, a daughter may gain knowledge about the elderly parent by engaging in sustained, physical nurturing. Through intimate engagement, the caregiver deciphers what the parent holds dear, and what counts as pain and comfort, or boredom and interest, to the parent. Her knowledge may be largely "an intuitive, nonverbal sort that is difficult to translate into clear proof."[32]

These ways of knowing catalyze moral problems for a more traditional model, because they vest those who hold such knowledge with responsibilities that a traditional model denies or even obstructs. The tendency to discount subjective awareness of patients' wishes and defer to a best interests standard is one example of such a problem. According to one group of commentators, few elderly patients have documented or made known their wishes concerning aggressive medical care, for example, by signing Living Wills or designating a durable power of attorney. "Nor can we easily infer from 'lifestyle' or vaguely expressed 'values' of the patient an acceptable course of action." The upshot is that, in most cases, the only appropriate standard will be one that focuses attention on the (objective) best interests of the patient."[33] Thus, the absence of clear proof about patients' preferences inclines these authors to defer to a more objective measure. Yet this approach effectively dismantles any effort to locate remnants of patients' values through intimate contacts and piece them into meaningful patterns. The result is that, lacking the patient who speaks clearly or leaves trails of objective evidence, the patient as person eludes us.

An alternative approach that heeds a subjective knowledge base would grant to families surrogate authority for their incompetent members, or would grant a rebuttable presumption in favor of family authority. As Rhoden notes, rather than creating a presumption where none exists, this approach simply replaces a present presumption with a more appropriate one:

> Courts have adopted medicine's strong presumption for treatment and required that patients overcome it. ...Judicial acceptance of this presumption silently incorporates the insidious notion that the doctor is rightfully in charge, authorized to treat without consent unless the family can persuade a court otherwise.[34]

More explicitly, the ethical considerations that support family authority in decisions for incompetent patients begin with the observation that the familiar categories of competent and incompetent camouflage a realm of patients' subjective experience. If this lived world can be understood, it provides important information for determining what forms of treatment are respectful of or consonant with patients' values. The argument proceeds with the claim that intimacy and patient contact provide a foothold for accessing the subjective experiences of incompetent patients. Moreover, families are a common locus for intimate relations. If so, then family members are *prima facie* entitled to make surrogate decisions for incompetent members.

This *prima facie* entitlement might be thought of as following from the family's more general right to privacy. Schoeman, for example, maintains that families possess a right to privacy that holds against society at large and entitles members to exclude others from scrutinizing obtrusions into family occurrences.[35] Thinking of families as possessing a right to privacy or autonomy serves multiple purposes: it safeguards an intimate sphere from third-party intrusions; it enables an intimate group to foster meaningful autonomy for members; and, in the case of vulnerable or compromised persons, it secures protection from decisions by third parties, such as courts, that are inclined to discount the patient's lived world and defer to objective standards.

In response, it might be argued that family members may harbor negative feelings or be emotionally distraught in other ways that color their assessment of treatment options. Such feelings may well hinder thoughtful decision making. Correspondingly, simply being a patient's relative does not guarantee feelings of love and care toward the patient. If not guaranteed, and if these feelings are the grounds for attributing surrogate authority to families, then the grounds for attributing it are undercut.

In reply to these objections, it can be said that neither point buttresses the argument for family autonomy. The argument I am defending is not premised on the assumption that family members possess capacities to make disinterested decisions or display purely positive feelings toward patients. To see that this is so, consider the more familiar case of *individual* autonomy. Autonomy entitles individuals to make decisions affecting them, despite the fact that they are immersed in, rather than distanced from, such decisions. Moreover, individuals' autonomy is not legitimately taken away even when their self-love founders. In a similar vein, the family's right to have a hand in health care decisions for incompetent patients is not contingent on objectivity or untainted love toward patients. Rather, it is the very absence of distance and objectivity that justifies the claim to be surrogate decision makers. Even when family members' love is coupled with feelings of jealously, envy, and spite, family autonomy does not fall by the wayside. Clearly, this does not mean that families are entitled to use members to vent hostile feelings or that family autonomy should go unchecked. Rather, it implies that filial relations carry moral force even if they are not made in heaven (which they never are).

However, what if intimacy is absent in parent–child relationships altogether, or what if a patient has a close family but finds his or her most intimate contacts outside it? For example, intimacy may be most potent in a friendship or with an unmarried partner. Alternatively, it may be felt most strongly in the daily care a nurse or therapist gives. Rather than challenging intimacy as a basis for surrogate authority, this objection challenges the contention that families are the locus of intimacy.

One important response to this objection is to acknowledge that intimacy, not filial status, is what ultimately provides the moral basis for surrogate authority. This is why the presumption in favor of family authority is presumptive *only.* To justify even a rebuttable presumption, however, we need to recognize how unlikely it is that the state or others will be able to start with no presumptions and go on to make an appropriate determination of where the patient's intimate contacts lie.

Another way of meeting this objection is to broaden the concept of the family to include less traditional associations. For example, New York's highest court recently expanded the definition of *family* as it applies to rent control laws to include "adults who show long-term financial and emotional commitment to each other even if they don't fit the traditional meaning of a family." The court majority reasoned that a definition of family "should not rest on fictitious legal distinctions or genetic history, but instead should find its foundation in the reality of family life...a more realistic, and certainly equally valid view of a family includes (persons) whose relationship is long-term and characterized by interdependence." The court's approach thus defies simple formulas and requires considering the "totality of the relationship as evidenced by the dedication, caring and self-sacrifice of the parties."[36]

Defining *family* in this way avoids the objection noted earlier to granting surrogate authority to families. Admittedly, it may pose other problems, such as the possibility of power struggles between traditional and nontraditional family members, but such conflicts can occur even among members related by blood or marriage. With this emendation, then, a *rebuttable* presumption in favor of family authority emerges as the best means for accomplishing informed surrogate decisions and safeguarding family autonomy.

The Family as a Commons

Intimate associations should not only ground decisional authority in the case of incompetent patients, but should also impose limits on how far respect for patients' wishes extends. This becomes clear in situations where elderly parents are competent to make health care choices, but their choices conflict with the preferences of offspring who serve as caregivers. I shall take it as a given that offspring are not entitled to ride roughshod over patients' preferences. Generally speaking, when families fail to safeguard members' interests at some threshold level, various forms of intervention are legitimate.

Consider the ways in which conflict cases often are treated in the bioethics literature. For example, many would agree that "external factors," that is, considerations that yield a benefit or burden to some party other than the patient, should not influence clinical decisions in any way.[37] What is more, "however sensitive the physician must be to the emotions and concerns of the family, he ought to remember that his covenant is with the patient, not the family."[38] Although a greater awareness of the stake of families in a treatment decision is shown by some,[39–42] consulting the family is generally deemed to be nothing more than a medical courtesy, and the family is thought to have neither ethical nor, in most states, legal authority to make treatment decisions.[43,44]

One difficulty with this standard approach is that considerations of justice limit the extent to which persons' autonomous choices can be satisfied. More often than not, the diverse goods individuals need to execute medical decisions are drawn from a shared pool. The crux of this argument against the more standard view is that justice is a family virtue, and that members have filial rights and duties. Just as autonomy and privacy are values for families as well as individuals, justice constitutes an ethical requirement in the family, as well as the larger society. Excluding these values from the family creates an ethical void, whereas designating them as family goods illuminates filial responsibilities.

One place where a sense of limits is gaining wider recognition is the federal level. In particular, the rising costs of Medicare and Medicaid are sparking a heated debate about the limits of individuals' rights to publicly funded medical care. Obviously, if each individual in need of an artificial heart or a liver transplant is granted one, joint resources and dollars would soon be exhausted. To preserve social goods and ensure respect for values that emerge at a social level, such as justice, individuals' wishes are not always sacrosanct.[45–49]

The debate about scarce medical resources makes evident, on the one hand, that, when persons depend on the resources of others, others participate implicitly in decisions to use these resources. They affect the range of options and the costs of pursuing them. On the other hand, this *de facto* role often remains hidden. For example, hospitals and staff develop informal protocols that distribute resources by ability to pay, or on a first-come-first-served basis, rather than implementing more overt and systematic strategies. We like to think of individuals as unbounded, free agents, yet, on reflection, the larger society has moral authority to choose how to use and distribute its common stock.

These considerations raise doubts about the prospect of viewing individual autonomy as an ethical absolute in the family setting as well. If the wishes of the geriatric patient usurp all other considerations, then others' claims to contribute to decisions affecting shared goods will be ignored. Just as the larger society is entitled to determine how its common goods will be dispersed, the family has moral authority to influence decisions that expend its resources and draw on its support.

As indicated, offspring frequently bear heavy emotional and financial burdens when family members become ill. They thus have separate and potentially conflicting interests in treatment decisions.

Greater recognition of these interests is appropriate. For example, if the needs of children who serve as caregivers are not accommodated when in conflict with a parent's desire to obtain an immediate hospital discharge, the childrens' lives are unduly disrupted. If no attention is paid by hospitals and staff to the financial wherewithal, emotional stamina, and unique strengths and weaknesses of family caregivers, and if all decisions are perfunctorily delegated to patients, the outcome may undermine family unity or compromise individual members. The paucity of public funding and support for home health care and the lack of viable long-term care alternatives for many already places caregivers in a vulnerable position. This gives added importance to limiting the demands placed on them and recognizing that they, as well as patients, are in need of supportive measures.

In these situations, the family functions like a commons.[50] A tragedy of the commons occurs if members seek to take more common goods than they are entitled to or refuse to make appropriate sacrifices and contributions. A commons flourishes only if each party is willing to acknowledge that resources are shared commodities and respects others' stakes in preserving them. Thus, even if competent patients formulate clear preferences, these preferences are not the final word on what constitutes an ethically sound decision. Ethically sound decision making must also consider the importance of supporting the commons on which patients and other family members depend. In the event that patients' preferences take no heed of the family commons, their preferences may be legitimately overridden, for example, to prevent a tragedy of the commons from occurring. It would be a mistake, however, to interpret this approach as entitling medical personnel to bypass competent patients altogether and go directly to family members. Clearly, any decision process that excludes a role for competent patients is indefensible. A presumption in favor of supporting the commons serves only to attenuate an exclusive autonomy focus. It establishes that individual autonomy is not an absolute or uncompromising value in medical decisions.

Of course, a tragedy of the commons can occur around any family member, young or old. It would be a mistake to suppose the contrary, that is, that only the old can pose a threat to the family commons. After all, an imperiled newborn is equally capable of draining family resources without even demanding them, and society may be more likely to idealize parents who spend down to poverty to pay for offsprings' graduate or professional educations than to idealize financial or professional sacrifices made for aging parents. With the growing number of older parents relying on family supports of various kinds, it is important to underscore that the young, especially infants and young children, are by far the largest drain on family resources.

One concern these remarks may elicit is that invoking a vocabulary of rights and duties in filial contexts will cause other goods, such as benevolence and generosity, to be displaced. Ruddick voices this concern, for example, when she states that "abstract rules of fairness badly serve the institutions that rely on them to structure personal relations."[51]

In response, it should be noted that the virtues displayed in personal relationships need not function as adversaries to justice. Nor need justice lay dormant when benevolence flourishes. For example, the even-handed sharing of parental energies between offspring displays justice, whereas the generosity of parental giving simultaneously reveals benevolence. Likewise, just parenting implies that each offspring deserves to have his or her viewpoints heard, rather than silenced; at the same time, nurturance and caring call forth attentive, supportive listening.

These examples illustrate how a moral point of view in personal relationships excludes partiality at one level, yet seeks to incorporate it at another. In the first example, a pattern of flagrantly favoring one child would wrongly convey that other offspring are less worthy of parental time and attention. Impartiality is thus appropriate, but so too is generous giving. The latter signals to offspring that they occupy a special place in their parents' lives. In the second example, an impartial viewpoint forbids ignoring an offspring's views altogether, even if summoning an attentive, caring response is not possible.

To claim a place for justice in the family and express the possibility of harmonizing it with other virtues is not to deny the real difficulty of balancing these diverse goods. Problems can arise in two important ways: intimacy may impede one's ability to become attuned to justice, or the process of detaching in order to consider justice may infect and destroy intimacy. As Thoreau puts it:

> I...am sensible of a certain doubleness by which I can stand on remote from myself as from another. However intense my experience, I am conscious of the presence and criticism of a part of me, which, as it were, is not part of me, but spectator, sharing no experience, but taking note of it. ...This doubleness may...make us poor neighbors and friends sometimes."[52]

Safeguarding against these occurrences requires cultivating an ability to adopt both partial and impartial viewpoints and alternate between them. This ability to shift viewpoints calls on our capacity to represent to ourselves the experience of being at once separated from and connected with others.[53,54]

Clearly, this objection poses more deep and difficult problems than can be considered here. Elsewhere, I have explored it in more detail.[55] At the very least, these remarks call into question the assumption this objection rests on, namely, that considerations of justice can have no place (or no central place) in close associations, such as families. Viewing the family as a commons can, in fact, be thought of as invoking a philosophical tradition with classical roots.[56] Historically, political philosophers have made much of the similarities between families and states, and derived their accounts of political obligation from a filial model. We should not be too surprised, then, if distributive justice pertains to families as well as states.

The considerations raised here are intended to recapture some of the moral complexity that can be lost when dominant ideas, such as individual autonomy, take hold. Fortunately, the clinical reality is that families are never ignored, and the tension that arises by taking the family's welfare into account is forcefully felt. If my reasoning is sound, this is as it should be.

References

[1]N. Rhoden (1988) Litigating life and death. *Harvard Law Review,* **102,** 375–446.

[2]R. Burt (1979) Conversation with silent patients, in *Taking Care of Strangers.* Free Press, New York, NY, p. 152, emphasis added.

[3]L. Blum (1980) *Friendship, Altruism, and Morality.* Routledge and Kegan Paul, London, UK.

[4]L. Blum (1987) Particularity and responsiveness, in *The Emergence of Morality in Young Children* (J. Kagan and S. Lamb, eds.), University of Chicago, Chicago, IL, pp. 306–336.

[5]S. Ruddick (1989) *Maternal Thinking.* Beacon Press, New York, NY.

[6]J. B. Schneewind (1986) The use of autonomy in ethical theory, in *Reconstructing Individualism* (T. C. Heller, M. Sosna, and D. E. Wellbery, eds.), Stanford University, Stanford, CA, pp. 64–75.

[7]C. Gilligan (1982) *In a Different Voice.* Harvard University, Cambridge, MA.

[8]C. Gilligan, J. V. Ward, and J. M. Taylor, eds. (1989) *Mapping the Moral Domain.* Harvard University, Cambridge, MA.

[9]E. F. Kittay and D. T. Meyers, eds. (1987) *Women and Moral Theory.* Rowman and Littlefield, Totowa, NJ.

[10]G. Graham and H. LaFollete, eds. (1989) *Person to Person.* Temple University, Philadelphia, PA.

[11]M. Pearsall, ed. (1986) *Women and Values.* Wadsworth, Belmont, CA.

[12]D. Meyers, C. Murphy, and K. Kipnis, eds. (in press) *Kindred Matters: Rethinking the Philosophy of the Family.* Cornell University Press, Ithaca, NY.

[13]R. D. Adelman, M. G. Greene, and R. Charon (1987) The physician–elderly patient–companion triad in the medical encounter. *The Gerontologist* **27,** 729–734.

[14]E. Shanas (1979). The family as a social support system in old age. *The Gerontologist* **19,** 169–178.

[15]V. G. Cicirelli (1983) Adult children and their elderly parents, in *Family Relationships in Later Life* (T. H. Brubaker, ed.), Sage, Beverly Hills, CA, pp. 31–46.

[16]M. Livingston (1987) How illness affects patients' families. *British Journal of Medicine* **38,** 51–53.

[17]E. Kay (1982) Untitled. *Hospital Update* **Feb.,** 161–170

[18]Anonymous (1983) Cancer care: The relative's view. *Lancet* **3,** 1188.

[19]Cicirelli, in *Family Relationships in Later Life.*

[20]E. P. Stroller (1983) Parental care-giving by adult children. *Journal of Marriage and Family* **45,** 851–858.

[21]D. Callahan (1988) Families as caregivers: The limit of morality. *Archives of Physical Medical Rehabilitation* **69,** 323–328.

[22]Adelman, Greene, and Charon, *The Gerontologist* **27.**

[23]I. Roscowe (1981) Coalition in geriatric medicine, in *Elderly Patients and Their Doctors* (M. R. Haug, ed.), Springer, New York, NY, pp. 137–146.

[24]S. H. Miles (1987) Futile feeding at the end of life: Family virtues and treatment decisions. *Theoretical Medicine,* **8,** 293–302.

[25]Roscowe, in *Elderly Patients and Their Doctors.*

[26]Ibid.

[27]H. Brody (1978) The role of the family in medical decisions. *Theoretical Medicine* **8,** 253–257.

[29]Rhoden, *Harvard Law Review,* **102,** 438–439.

[30]S. Gadow (1980) Existential advocacy: Philosophical foundation in nursing, in *Nursing Images and Ideals* (S. F. Spicker and S. Gadow, eds.), Springer, New York, NY, pp. 79–101.

[31]M. H. Waymack and G. A. Taler (1988) *Medical Ethics and the Elderly,* Pluribus, Chicago, IL.

[32]Rhoden, *Harvard Law Review,* **102,** 392.

[33]R. Sherlock and M. Dingus (1985) Families and the gravely ill: Roles, rules, and rights. *Journal of the American Geriatric Society* **33,** 121–124.

[34]Rhoden, *Harvard Law Review,* **102,** 379.

[35]F. Schoeman (1980) Rights of children, rights of parents, and the moral basis of the family. *Ethics* **91,** 6–19.

[36]P. S. Gutis (1989) Court widens family definition to gay couples living together. *New York Times,* July 7, pp. 1,12.

[37]M. Siegler (1982) Decision-making strategy for clinical ethical problems in medicine. *Archives of Internal Medicine* **142,** 2178,2179.

[38]Sherlock and Dingus, *Journal of the American Geriatric Society* **33,** 123.

[39]President's Commission for the Study of Ethical Problems in Medicine and Biomedical and Behavioral Research (1982) *Making Health Care Decisions.* US Government Printing Office, Washington, DC.

[40]Miles, *Theoretical Medicine* **8.**

[41]D. M. High and H. B. Turner (1987) Surrogate decision-making: The elderly's familial expectations. *Theoretical Medicine* **8,** 303–320.

[42]M. Yarborough (1988) Continued treatment of the fatally ill for the benefit of others. *Journal of the American Geriatrics Society* **36,** 63–67.

[43]A. R. Jonsen, M. Siegler, and W. J. Winslade (1988) *Clinical Ethics,* 2nd ed., Macmillan, New York, NY.

[44]J. Areen (1987) The legal status of consent obtained from families of adult patients to withhold or withdraw treatment. *JAMA* **258,** 229–235.

[45]N. S. Jecker (1989a) Should we ration health care? *Journal of Medical Humanities and Bioethics* **10,** 77–90.

[46]N. S. Jecker and R. A. Pearlman (1989) Ethical constraints on rationing medical care by age. *Journal of the American Geriatric Society* **37,** 1067–1075.

[47]P. A. L. Haber (1986) Rationing is reality. *Journal of the American Geriatric Society* **34,** 761–763.

[48]V. R. Fuchs (1984) The "rationing" of medical care. *New England Journal of Medicine* **311,** 1572,1573.

[49]R. W. Evans (1983) Health care technology and the inevitability of resource allocation and rationing decisions I. *JAMA* **249,** 2047–2053.

[50]G. Hardin (1983) Living on a lifeboat, in *Moral Issues* (J. Narveson, ed.), Oxford University, New York, NY, pp. 167–178.

[51]Ruddick, Fostering growth, in *Maternal Thinking,* p. 97

[52]H. D. Thoreau (1960) Walden, in *Walden and Civil Disobedience* (S. Paul, ed.), Houghton, Mifflin, Boston, MA, pp. 1–227.

[53]C. Gilligan (1986) Remapping the moral domain, in *Reconstructing Individualism* (T. C. Heller, M. Sosna, and D. E. Wellbery, eds.), Stanford University, Stanford, CA, pp. 237–252.

[54]N. P. Lyons (1983) Two perspectives: On self, relationships and morality. *Harvard Educational Review* **53,** 125–145.

[55]N. S. Jecker (in press) Impartiality and special relations, in *Kindred Matters: Rethinking the Philosophy of the Family* (D. T. Meyers, C. Murphy, and K. Kipnis, eds.), Cornell University Press, Ithaca, NY.

[56]N. S. Jecker (1989b) Are filial duties unfounded? *American Philosophical Quarterly* **26,** 73–80.

Distributive Justice in an Aging Society

Limiting Health Care for the Old

Daniel Callahan

In October 1986, Dr. Thomas Starzl of Presbyterian University Hospital in Pittsburgh successfully transplanted a liver into a 76-year-old woman, thereby extending to the elderly patient the most technologically sophisticated and expensive kind of medical treatment available (the typical cost of such an option is more than $200,000). Not long after that, Congress brought organ transplants under Medicare coverage, thus guaranteeing an even greater range of this form of lifesaving care for older age groups.

That is, on its face, the kind of medical progress we usually hail: a triumph of medical technology and a newfound benefit provided by an established health care program. However, at the same time those events were taking place, a government campaign for cost containment was under way, with a special focus on health care to the aged under Medicare. It is not hard to understand why. In 1980, people over age 65—11% of the population—accounted for 29% of the total American health care expenditures of $219.4 billion. By 1986, the elderly accounted for 31% of the total expenditures of $450 billion. Annual Medicare costs are projected to rise from $75 billion in 1986 to $114 billion by the year 2000, and that is in current, not inflated, dollars.

Is it sensible, in the face of the rapidly increasing burden of the health care costs for the elderly, to press forward with new and expensive ways of extending their lives? Is it possible even to hope

Aging and Ethics Ed.: N. Jecker

to control costs while simultaneously supporting innovative research, which generates new ways to spend money? Those are now unavoidable questions. Medicare costs rise at an extraordinary pace, fueled by an increasing number and proportion of the elderly. The fastest-growing age group in the United States is comprised of those over age 85, increasing at a rate of about 10% every two years. By the year 2040, it has been projected, the elderly will represent 21% of the population and consume 45% of all health care expenditures. How can costs of that magnitude be borne?

Anyone who works closely with the elderly recognizes that the present Medicare and Medicaid programs are grossly inadequate in meeting their real and full needs. The system fails most notably in providing decent long-term care and medical care that does not constitute a heavy out-of-pocket drain. Members of minority groups and single or widowed women are particularly disadvantaged. How will it be possible, then, to provide the growing number of elderly with even present levels of care, much less to rid the system of its inadequacies and inequities, and, at the same time, add expensive new technologies?

The straight answer is that it will be impossible to do all those things and, worse still, it may be harmful even to try. It may be so because of the economic burden that would be imposed on younger age groups, and because of the requisite skewing of national priorities too heavily toward health care. However, that suggests to both young and old that the key to a happy old age is good health care, which may not be true.

In the past few years, three additional concerns about health care for the aged have surfaced. First, an increasingly large share of health care is going to the elderly rather than to youth. The Federal government, for instance, spends six times as much providing health benefits and other social services to those over 65 as it does to those under 18. Also, as the demographer Samuel Preston observed in a provocative address to the Population Association of America in 1984, "Transfers from the working-age population to the elderly are also transfers away from children, since the working ages bear far more responsibility for childrearing than do the elderly."

Preston's address had an immediate impact. The mainline senior citizen advocacy groups accused Preston of fomenting a war between the generations, but the speech also stimulated Minnesota Senator David Durenberger and others to found Americans for Generational Equity (AGE) to promote debate about the burden on future generations, particularly the Baby Boom cohort, of "our major social insurance programs." Preston's speech and the founding of AGE signaled the outbreak of a struggle over what has come to be called "intergenerational equity," which is now gaining momentum.

The second concern is that the elderly, in dying, consume a disproportionate share of health care costs. Stanford University economist Victor Fuchs notes:

> At present, the United States spends about 1 percent of the gross national product on health care for elderly persons who are in their last year of life. ...One of the biggest challenges facing policy makers for the rest of this century will be how to strike an appropriate balance between care for the [elderly] dying and health services for the rest of the population.

The third issue is summed up in an observation by Dr. Jerome Avorn of the Harvard Medical School, who wrote in *Daedalus,* "With the exception of the birth-control pill, [most] of the medical-technology interventions developed since the 1950s have their most widespread impact on people who are past their fifties—the further past their fifties, the greater the impact." Many of the techniques in question were not intended for use on the elderly. Kidney dialysis, for example, was developed for those between the ages of 15 and 45. Now some 30% of its recipients are over 65.

The validity of those concerns has been vigorously challenged, as has the more general assertion that some form of rationing of health care for the elderly might become necessary. To the charge that old people receive a disproportionate share of resources, the response has been that assistance to them helps every age group: It relieves the young of the burden of care that would otherwise have to bear for elderly parents and, since those young will eventually become old, promises them similar care when they need it. There is no guarantee, moreover, that any cutback in health care for the elderly would result in a transfer of the savings directly to the young. Also, some ask, "Why should we contemplate restricting care for the elderly when we wastefully spend hundreds of millions on an inflated defense budget?"

The assertion that too large a share of funds goes to extending the lives of elderly people who are terminally ill hardly proves that it is an unjust or unreasonable amount. They are, after all, the most in need. As some important studies have shown, it is exceedingly difficult to know that someone is dying; the most expensive patients, it turns out, are those who were expected to live but died. That most new technologies benefit the old more than the young is logical: most of the killer diseases of the young have now been conquered.

There is little incentive for politicians to think about, much less talk about, limits on health care for the aged. As John Rother, director of legislation for the American Association of Retired Persons, has observed, "I think anyone who wasn't a champion of the aged is no longer in Congress." Perhaps also, as Guido Calabresi, dean of the Yale Law School, and his colleague Philip Bobbitt observed in their thoughtful 1978 book *Tragic Choices,* when we are forced to make painful allocation choices, "Evasion, disguise, temporizing...[and] averting our eyes enables us to save some lives even when we will not save all."

I believe that we must face this highly troubling issue. Rationing of health care under Medicare is already a fact of life, though rarely labeled as such. The requirement that Medicare recipients pay the first $520 of hospital care costs, the cutoff of reimbursement for care after 60 days, and the failure to cover long-term care are nothing other than allocation and cost-saving devices. As sensitive as it was to the senior citizen vote, the Reagan Administration agreed only grudgingly to support catastrophic health care coverage for the elderly (a benefit that will not help very many of them), and it expressed its opposition to the version of the bill recently passed by the house. Any administration is bound to be far more resistant to long-term health care coverage.

However, there are reasons other than the economics to think about health care for the elderly. The coming economic crisis provides a much needed opportunity to ask some deeper questions. Just what is it that we want medicine to do for us as we age? Other cultures have believed that aging should be accepted, and that it should be, in part, a time of preparation for death. Our culture seems increasingly to dispute that view, preferring instead, it often seems, to think of aging as hardly more than another disease, to be fought and rejected. Which view is correct?

Let me interject my own opinion. The future goal of medical science should be to improve the quality of old people's lives, not to lengthen them. In its long-standing ambition to forestall death, medicine has reached its last frontier in the care of the aged. Of course, children and young adults still die of maladies that are open to potential cure, but the highest proportion of the dying (70%) are over 65. If death is ever to be humbled, that is where endless work remains to be done, but however tempting the challenge of that last frontier, medicine should restrain itself. To do otherwise would mean neglecting the needs of other age groups and of the old themselves.

Our culture has worked hard to redefine old age as a time of liberation, not decline, a time of travel, of new ventures in education and self-discovery, of the ever-accessible tennis court or golf course, and of delightfully periodic but thankfully brief visits from well behaved grandchildren. That is, to be sure, an idealized picture, but it arouses hopes that spur medicine to wage an aggressive war against the infirmities of old age. As we have seen, the costs of such a war would be prohibitive. No matter how much is spent, the ultimate problem will still remain: people will grow old and die. Worse still, by pretending that old age can be turned into a kind of endless middle age, we rob it of meaning and significance for the elderly.

There is a plausible alternative: a fresh vision of what it means to live a decently long and adequate life, what might be called a "natural lifespan." Earlier generations accepted the idea that there was a natural lifespan—the biblical norm of three score and ten captures that notion (even though, in fact, that was a much longer lifespan than was typical in ancient times). It is an idea well worth reconsidering, and would provide us with a meaningful and realizable goal. Modern medicine and biology have done much, however, to wean us from that kind of thinking. They have insinuated the belief that the average lifespan is not a natural fact at all, but instead one that is strictly dependent on the state of medical knowledge and skill. Also, there is much to that belief as a statistical fact: The average life expectancy continues to increase, with no end in sight.

However, that is not what I think we ought to mean by a natural lifespan. We need a notion of a full life that is based on some deeper understanding of human needs and possibilities, not on the state of medical technology or its potential. We should think of a natural lifespan as the achievement of a life that is sufficiently long to take advantage of those opportunities life typically offers and that we ordinarily regard as its prime benefits—loving and "living," raising a family, engaging in work that is satisfying, reading, thinking, cherishing our friends and families. People differ on what might be a full natural lifespan; my view is that it can be achieved by the late 70s or early 80s.

A longer life does not guarantee a better life. No matter how long medicine enables people to live, death at any time—at age 90 or 100 or 110—would frustrate some possibility, some as-yet-unrealized goal. The easily preventable death of a young child is an outrage. Death from an incurable disease of someone in the prime of young adulthood is a tragedy. However, death at an old age, after a long and full life, is simply sad, but it is a part of life itself.

As it confronts aging, medicine should have as its specific goals the averting of premature death, that is, death prior to the completion of a natural lifespan, and thereafter, the relief of suffering. It should pursue those goals so that the elderly can finish out their years with as little needless pain as possible—and with as much vitality as can be generated in contributing to the welfare of younger age groups and to the community of which they are a part. Above all, the elderly need to have a sense of the meaning and significance of their stage in life, one that is not dependent on economic productivity or physical vigor.

What would medicine oriented toward the relief of suffering rather than the deliberate extension of life be like? We do not have a clear answer to that question, so long standing, central, and persistent has been medicine's preoccupation with the struggle against death. However, the hospice movement is providing us with much guidance. It has learned how to distinguish between the relief of suffering and the lengthening of life. Greater control by elderly persons over their own dying—and particularly an enforceable right to refuse aggressive life-extending treatment—is a minimal goal.

What does this have to do with the rising cost of health care for the elderly? Everything. The indefinite extension of life combined with an insatiable ambition to improve the health of the elderly is a recipe for monomania and bottomless spending. It fails to put health in its proper place as only one among many human goods. It fails to accept aging and death as part of the human condition. It fails to present to younger generations a model of wise stewardship.

How might we devise a plan to limit the costs of health care for the aged under public entitlement programs that is fair, humane, and sensitive to their special requirements and dignity? Let me suggest three principles to undergird a quest for limits. First, government has a duty, based an our collective social obligations, to help people live out a natural lifespan, but not to help medically extend life beyond that point. Second, government is obliged to develop under its research subsidies, and to pay for, under its entitlement programs, only the kind and degree of life-extending technology necessary for medicine to achieve and serve the aim of a natural lifespan. Third, beyond the point of a natural lifespan, government should provide only the means necessary for the relief of suffering, not those for life-extending technology.

A system based on those principles would not immediately bring down the cost of care of the elderly; it would add cost, but it would set in place the beginning of a new understanding of old age, one that would admit of eventual stabilization and limits. The elderly will not be served by a belief that only a lack of resources, better financing mechanisms, or political power stands between them and the limitations of their bodies. The good of younger age groups will not be served by inspiring in them a desire to live to an old age that maintains the vitality of youth indefinitely, as if old age were nothing but a sign that medicine has failed in its mission. The future of our society will not be served by allowing expenditures on health care for the elderly to escalate endlessly and uncontrollably, fueled by the false altruistic belief that anything less is to deny the elderly their dignity. Nor will it be aided by the pervasive kind of self-serving argument that urges the young to support such a crusade because they will eventually benefit from it also.

We require, instead, an understanding of the process of aging and death that looks to our obligation to the young and to the future, that recognizes the necessity of limits and the acceptance of decline and death, and that values the old for their age and not for their continuing youthful vitality. In the name of accepting the elderly and repudiating discrimination against them, we have succeeded mainly in pretending that, with enough will and money, the unpleasant part of old age can be abolished. In the name of medical progress, we have carried out a relentless war against death and decline, failing to ask in any probing way if that will give us a better society for all.

A Lifespan Approach to Health Care

Norman Daniels[1]

Conflicting Messages About Our Health Care System

The next few years will bring many proposals intended to reform the financing and the design of our health care system. Major reforms are necessary because we simultaneously face several critical issues: a growing insurance gap now involving nearly 40 million people; rapidly rising health care costs whose long-term rate of increase is unaffected by current cost-containment measures; and a pattern of resource allocation that leaves many fundamental health care needs unmet. In thinking about what we want reforms to accomplish, I think we should try to answer such questions as these: Do we get from the spectrum of care we receive over the lifespan what we need and most want? Does our health care system allocate resources over the whole lifespan in a way that is fair to all age groups? How should we as citizens and as health professionals think about what we want our health care system to do for us? What are our social obligations to design a system that delivers care to both rich and poor, black and white, male and female, young and old? What does justice require in the way of health care over the lifespan?

These are hard questions, and it is no surprise that our society has not faced them as squarely and directly as it should have. Part of what makes answering them so difficult is the very different

Aging and Ethics Ed.: N. Jecker

messages we, the public, get about our health care system. One message we get is that *medical technology can do just about anything, at all points in the lifespan, given the resources.* The media dramatize the latest ventures in in vitro fertilization, the microsurgery on minute neonates, and the saving of children with biliary atresia by liver transplants. A young worker who loses a hand in an industrial accident can have it reattached; genetic engineering gives us new drugs to treat heart attacks; whole hip sockets can be replaced for the elderly. We are experimenting with transplanting neural tissue to repair spinal damage and even to treat Alzheimer's and Parkinson's diseases. Cosmetic surgeons are vying with Henry Moore as sculptors of fatty tissue. We use technology in confusion and desperation: we can keep alive through heroic measures those who have lost central features of their personhood, sometimes in disregard of their wishes. The message here is that from the marriage bed to the deathbed we can hold death, disability, and ugliness at bay if we are willing.

There are darker messages, too. We read that 40 million Americans have no health care insurance at all and that we are the only industrial democracy that fails to provide universal access to basic health care, let alone these technological miracles. We read that, when prenatal maternal care programs were eliminated from federal funding as a result of budget-cutting in the early Reagan years, infant mortality rates rose—at the same time we invested heavily in neonatal intensive care units. We read that millions of partially disabled elderly are unable to get adequate home care or social support services. We read that millions of mentally ill people are unable to get treatment at all—and they live in our streets among the homeless.

Yet, despite all these things that we fail to do—thereby "saving" health care dollars, we also read that health care costs are rising at rates above inflation. This rise takes place despite federal DRG capping of Medicare hospital payments, increased cost-sharing by Medicare patients, decreased tax deductions for medical care, state capping of hospital budgets, and measures taken by employers and third-party payers intended to promote price competition in health care or prospective payment wherever possible. Because these measures have not greatly slowed rising costs, and because some economists do not think we can succeed just by measures that try to trim the waste or fat out of the system—indeed, this is the central dispute among the experts—we hear more and more calls for the *rationing* of medical care.

In fact, of course, we already ration beneficial care by ability to pay, for that is what leaving 40 million people without insurance coverage amounts to. Similarly, some beneficial care (and not just unnecessary care) is denied patients as a result of DRGs and other cost-containment measures, and some studies have shown a slight increase in death rates for the elderly as a result of shortened hospital stays following surgery. However, the new call for rationing is a request for more explicit measures. It is a call for us to make decisions about the priorities that should govern the dissemination of medical technologies.

This call for rationing in the United States is striking. We are not in wartime. We do not require national sacrifice to keep our soldiers in battle. Can we not have both guns and bandages? Can we not give up some guns for some bandages? This is not a depression. Productivity is not so low that we cannot keep Reeboks on our feet and a chicken in every microwave. Yet, what many are saying is that spending 11 or 12% of GNP on health care may be approaching the limit of what it is reasonable to invest in health care, and that our current cost-containment measures can at best squeeze a small amount of fat out of the system. Also, they say, quite plausibly, that if we do not figure out what kinds of medical technologies are most important to disseminate, then our costs will continue to rise rapidly, even though we will only achieve marginal gains in health status. What is worse, there is great reluctance to introduce entitlement schemes that eliminate the shameful gaps in health care insurance without some demonstration that we know how to control health care costs, through rationing or otherwise. In effect, we are being told that the price for making health care more accessible to all is that we may have to make some forms of it less accessible to all—and this means we will have to face hard policy choices.

There is much truth in all of these messages about health care: we do amazing things; we fail to do equally important things; we will have to refrain from doing other important things. It will be difficult to expand entitlement to health care without controlling health care costs, but controlling costs involves our making hard choices about what technologies to use. When not everyone can get what he or she wants or needs, we must find principles of justice or fair procedures for determining who should get what, but this kind of talk about competition for scarce resources has its political costs, and I would like to say a few things about them.

The Elderly in the Hot Seat

When public talk turns to claims about scarcity and the need for rationing, there is a tendency to blame one group for the problems of another. In recent years, there has been considerable finger-pointing at the elderly—not by the public as a whole, which continues to support programs aimed at the needs of the elderly, but at least by some scholars, planners, and even legislators. What has emerged is a growing perception that the old and the young are locked in fierce competition for critical but scarce resources, namely, public funds for human services. It is in this regard that we hear complaints that the old are benefiting at the direct expense of children, of the poor, and of younger workers. It is in this context that some call for "intergenerational equity"—indeed, there is a Washington lobby called AGE (Americans for Generational Equality).

The most plaintive cries are about competition between the elderly and children—between grandparent or great-grandparents and grandchildren.

> Since 1970, expanded Social Security benefits have reduced poverty among the elderly from double the national incidence to a level slightly below the average rate of poverty. There

are now proportionally more poor children than poor elderly, a switch from 15 years ago.

- Children now receive a smaller proportion of Medicaid dollar than in earlier years, despite an increase in the number of poor children.
- In 1971, we spent less on the elderly than on national defense, but in the 1980s we spent more.
- Federal per capita expenditures on children are only nine percent of per capita expenditures on the elderly.

More generally, we spend four-and-a-half times as much on federal retirement programs, including Medicare, which benefit rich and poor alike, than we do on all welfare programs aimed specifically at the poor.

The elderly are thus seen as competitors with children and the poor. However, they are also portrayed as competitors with other birth cohorts: each retiree is now supported by 3.4 workers, but when the Baby Boom retires, only two workers will contribute for every retiree.

The conflict between the elderly and other groups over public money sometimes penetrates right to the level of the family. Currently, families provide about 80% of all home health care to the partially disabled elderly. This care is costly to adult children and other family members, in terms of money, expended time, and the stress of sustaining care over extended periods. As one daughter put it:

> We put her in the shower and brush her teeth. ...We put lipstick on her and put her hair in a french twist with floral combs. ...When we go away overnight, we take her with us. ...I put paper on the floor 'cause she doesn't always make it to the bathroom. ...I'll paint her apartment, change the light bulbs, and wash the windows.[2]

Despite these high levels of care provided by some families, rising public expenditure on long-term care have led some legislators in the early 1980s to propose "family responsibility initiatives" in about half the states. If passed, such laws would hold family members legally responsible for costs currently paid by Medicaid. Shifting costs out of public budgets, however, will not eliminate competition between the elderly and the young for resources. It will only shift the locus and burden of that competition from public budgets to family budgets.

Though we spend heavily—and some say too heavily—to meet the needs of the elderly, their needs are far from met by existing institutions. Perhaps the clearest example of the failure to meet important needs is our long-term care system. This system forces the premature institutionalization of many partially disabled elderly and leaves millions of others without adequate home care services to enable them to function without extreme hardship. Families that provide care get little relief from their burdens. Yet it is this already inadequate system that will be most strained by the rapid growth of age groups over 75. At the same time, we continue to provide intensive acute care treatments to the dying elderly. We point to this glamorous challenge to death as if it proved we value highly every last minute of their lives—and indeed it is the last minute we appear to value the most, though we do not always know which that minute will be. Yet it is far from obvious that prolonging the process of dying in these ways meets an important health care need. We often trap the elderly in treatments they and their families do not want, even ignoring explicit preferences to discontinue treatment. Children often insist on treating their parents more aggressively than they would want to be treated themselves.

I already noted that the proportion of the elderly who are poor is comparable to that of the population as a whole, which represents great progress. However, this still means that millions of elderly live in poverty and millions more live near the poverty line. Moreover, this poverty is concentrated by race and sex. We can reduce expenditures to meet the needs of the elderly only by pushing many more into poverty or by excluding them from medical care.

The problems we face are especially difficult because it is the basic needs of different groups that lead to conflict and competition. The old and the young both need health care. What do we do if competition means a choice between health care for the elderly and education for their grandchildren, or between the immediate need of elderly parents and the future needs of their adult children?

Underlying the common perception of competition between the old and the young, underlying the call for "generational equity" in our aging society, there lurks a challenging new problem: what is a just or fair distribution of social resources among the different age groups competing for them? I shall argue that we can solve this problem only if we think about the competition among age groups in a radically different way. Indeed, we will have to stop thinking about competition between groups altogether.

Justice and Health Care over the Lifespan

I have given an overview of the problems we face in our health care system because I want to emphasize one central fact: the time for piecemeal thought about individual problems and merely incremental modifications of the system is past. Of course, we can continue to improve it with legislation aimed at meeting specific needs, for example, by providing for long-term care needs of the elderly or by closing the insurance gap. However, we really face a much bigger challenge: we must rethink what we want out of our health care investment, and, in a principled way, we must consider what our social obligations are in the allocation of resources and the provision of access to care. We have never squarely faced these big tasks, at least not in the way other societies, such as Canada, have when they designed systems that guarantee universal access to adequate health care and provide for regional planning about the dissemination of technology and the allocation of health care resources.

Concerns about these larger issues are already surfacing at the state level. Massachusetts has passed legislation intended to guarantee access to care to all in the state, but there are serious questions whether it has developed an adequate plan to control costs.[3] Oregon has attracted national attention with its decision to provide no state funding for organ transplants for the poor. It argues that prenatal maternal outreach programs will save twice as many lives per dollar spent as such transplant programs, and its senate president, a physician, has argued that, as long as many citizens are without access to any insurance, it is wrong to spend enormous sums to provide benefits to but a few. Of course, by denying access to expensive high-technology treatments that may be available to other citizens or to those who move to another state, or that may be available if pleas for charity funds are successful, Oregon risks creating other inequities that are highly problematic and very visible.[4] Nevertheless, my point is that these problems of access and resource allocation must be faced everywhere and must be based on an attempt to figure out what justice requires. Massachusetts and Oregon are at least trying to face the right issues.

To address these issues of access and resource allocation, we must raise some basic questions about what justice requires in the design of our health care system.[5] Let us begin with a very general question: Is health care *special?* Is it social good that we should distinguish from other goods, say video recorders, because of its special importance? Does it have special moral importance? Also, does that moral importance mean that there are social obligations to distribute it in ways that might not coincide with the results of market distribution? I believe the answer to all these questions is "yes."

Health care—I mean the term quite broadly—does many important things for people. Some extends lives, some reduces pain and suffering, some merely gives important information about one's condition, and much health care affects the quality of life in other ways. Yet, we do not think all things that improve quality of life are comparable in importance: the *way* quality is improved seems critical. I have argued elsewhere that a central unifying function of health care is to maintain and restore functioning that is typical or normal for our species. Health care derives its moral importance from the following fact: normal functioning has a central effect on the opportunity open to an individual. It helps guarantee individuals a fair chance to enjoy the normal opportunity range for their society. The *normal opportunity range* for a given society is the array of life plans reasonable persons in it are likely to construct for themselves. An individual's fair share of the normal opportunity range is the array of life plans he or she may reasonably choose, given his or her talents and skills. Disease and disability shrinks that share from what is fair; health care protects it. Health care lets a person enjoy that portion of the normal range to which his or her full range of skill and talents would give him or her access assuming these too are not impaired by special social disadvantages. The suggestion that emerges from this account is that we should use impairment of the normal opportunity range as a fairly crude measure of the relative moral importance of health care needs at the macro level.

Some general theories of justice, most notably Rawls',[6] provide foundations for a principle protecting fair equality of opportunity. If such a principle is indeed a requirement of an acceptable general theory of justice, then I believe we have a natural way to extend such general theories to govern the distribution of health care. We should include health care institutions among those basic institutions of a society that are governed by the fair equality of opportunity principle.[7] If this approach to a theory of just health care is correct, it means that there are social obligations to provide health care services that protect and restore normal functioning. In short, the principle of justice that should govern the design of health care institutions is a principle that calls for guaranteeing fair equality of opportunity.

This principle of justice has implications for both access and resource allocation. It implies that there should be no financial, geographical, or discriminatory barriers to a level of care that promotes normal functioning. It also implies that resources be allocated in ways that are effective in promoting normal functioning. That is, since we can use the effect on normal opportunity range as a crude way of ranking the moral importance of health care services, we can guide hard public policy choices about which services are more important to provide. Thus, the principle does not imply that every technology that might have a positive impact on normal functioning for some individuals should be introduced: we must weigh new technologies against alternatives to judge the overall impact of introducing them on fair equality of opportunity—this gives a slightly new sense to the term "opportunity cost." The point is that social obligations to provide just health care must be met within the conditions of moderate scarcity that we face. This is not an approach that gives individuals a basic right to have all their health care needs met. There are social obligations to provide individuals only with those services that are part of the design of a system that, on the whole, protects equal opportunity.

We must refine this account so that it applies more directly to the problem of allocating health care over the lifespan—among different age groups. I draw on three basic observations. First, there is the banal fact we have all noticed: we age. By contrast, we do not change sex or race. This contrast has important implications for the problem of equality. If I treat blacks and whites or men and women differently, than I produce an inequality, and such inequalities raise questions about justice. If I treat the old and the young differently, I may or may not produce an inequality. If I treat them differently just occasionally and arbitrarily, then I will treat different persons unequally, but if I treat as a matter of policy the old one way and the young another, and I do so over their whole lives, then I treat all persons the same way. No inequality is produced. Thus, the fact that we all notice, that we age, means age is different from race or sex when we think about distributive justice.

Second, as we age, we pass through institutions that redistribute wealth and income in a way that performs a "savings" function. The observation is trivial with regard to income support institutions, such as the Social Security system. It is not often noticed that our health care system does the same thing. When we reach age 65, we consume health care resources at about 3.5 times the rate (in dollars) that we do prior to age 65. However, we pay, as working people, a combined health care insurance premium—through private premiums, through employee contributions, and through Social Security taxes—that covers not just our actuarially fair costs, but the costs of the elderly and of children as well. If this system continues as we age, others will pay "inflated premiums" that will cover our higher costs when we are elderly. In effect, the system allows us to defer the use of resources from one stage of our lives to a later one. It "saves" health care for our old age—when we need more of it.

Third, our health care system is not prudently designed, given that it plays this role as a savings institution. It lavishes life-extending resources on us as we are dying, but it withholds other kinds of services, such as personal care and social support services, which may be crucial to our well-being when our lives are not under immediate threat. The system could be far more prudently designed. It could pay better attention to matching services to needs at different stages of our lives and, thus, be more effective in its savings function.

Earlier, I claimed that the just design of our health care institutions should rest on a principle protecting fair equality of opportunity. Imagine that each of us has a lifetime allocation of health care services, which we can claim only if the appropriate needs arise, as a result of appealing to such a principle. Our task now is to allocate that fair share over the lifespan—and to do so prudently. In this exercise, we will find out what is just or fair between age groups by discovering what it is prudent to do between stages of life, over the whole lifespan. One way to make sure we do not bias this allocation, favoring one stage of life and, thus, one age group over another is to pretend that we do not know how old we are. We must allocate these resources imagining that we must live our whole lives with the result of our choices. One way we would refine our earlier principle of justice is to conclude that we should protect our fair shares of the normal opportunity range at each stage of life. Since we must live through each stage, we will not treat any one stage as less important than another.[8]

Notice what this rather abstract perspective—I call it the Prudential Lifespan Account—accomplishes: it tells us that we should not think of age groups as competing with each other, but as sharing a whole life. We want to make that life go as well as possible, and we must therefore make the appropriate decisions about what needs it is most important to meet at each stage of life. If we do this prudently, we will learn how it is fair to treat each age group. Instead of focusing on competition, we have a unifying perspective or vision. I am suggesting that, as individuals and as a society, we must think through the decisions we must make about our health care system from this perspective.

Implications for Resource Allocation

There are important implications of this perspective. How would prudent deliberators view the importance of various personal care and social support services for the partially disabled as compared to personal medical services? From the perspective of these deliberators, both types of care would have the same rationale and the same general importance. Personal medical services restore normal functioning and, thus, have a great impact on an individual's share of the normal opportunity range at each stage of life. However, so too do personal care and support services for the partially disabled. They compensate for losses of normal functioning in ways that enhance individual opportunity. It is not prudent to design a system such as ours, which ignores these health care needs, since they affect such a substantial portion of the later stages of life. If we pay attention only to acute care needs, than we are "saving" the wrong kinds of resources, or not enough resources.

A major criticism of the US health care system—that it encourages premature and inappropriate institutionalization of the elderly—should be assessed in this light. The issue becomes not just one of costs and the relative cost-effectiveness of institutionalization vs home care. Rather, opportunity range for many disabled persons will be enhanced if they are helped to function normally outside institutions. They will have more opportunity to complete projects and pursue relationships of great importance to them, or even to modify the remaining stages of their plans of life. Often, this issue is discussed in terms of the loss of dignity and self-respect that accompanies premature institutionalization or inappropriate levels of care. The underlying issue, however, is loss of opportunity range, which obviously has an effect on autonomy, dignity, and self-respect. Viewed in this light, the British and Canadian systems (which are quite different from each other), in which extensive home care services exist (at least in some Canadian provinces), far more respect the importance of normal opportunity range for the elderly than does our system. They put more of their resources into improving opportunity range for the substantial number of elderly who are disabled over significant periods of the late stages of life. We put our resources into marginally extending life when it is threatened in old age by actute episodes. I am suggesting their approach may be more prudent, because it better protects age-relative opportunity range than ours.

There is another implication of this approach that is quite controversial. Under certain resource constraints, prudent deliberators would prefer a distributive scheme that improves their chances of reaching a normal lifespan (normal life expectancy) to one that gives them a reduced chance of reaching normal lifespan, but a greater chance to live an extended span once normal life expectancy is reached. It would be prudent, in other words, to put more resources into reducing early death than into adding years very late in life. Under some conditions of resource scarcity, this might imply rationing some life-extending technologies by age, but this implication very much depends on how the scarcity works in the society.[9] In contrast to Callahan,[10] I am not advocating such rationing as general policy, either because it would help contain costs or because it would add meaning to our old age, as Callahan would ironically have it. In contrast to Callahan, I do not think there is only one way to add meaning to old age. My argument leads only to a very modest conclusion: under certain conditions of scarcity, rationing by age would not be impermissible and could be the fair way to allocate scarce resources. Of course, such rationing would have to meet other stringent conditions, including that the policy and its rationale be public and democratically selected. Nevertheless, there are far more important things we can do to make our current system both more effective and more just in its use of resources than introducing rationing by age, even for those cases where scarcity may have the effects required by the argument I have sketched.

There are important practical implications of the Prudential Lifespan Account for the young as well as the old. The most basic implication of this view is that the insurance gap that excludes about 1 in 5 children from any form of medical insurance is simply intolerable. Important groups are addressing this problem in the traditional American way—patch as patch can. The Academy of Pediatrics is supporting legislation calling for universal health coverage for children and mothers—mirroring Medicare's protection at the end of life. Similarly, the Children's Defense Fund calls for expansion of Medicaid eligibility requirements (covering all children in families living below 200% of poverty). These proposals are consistent with the approach to justice I have been describing—provided they really solve the problems of access. There are good reasons now to consider solving this problem in an unAmerican way—through a truly universal insurance scheme. Incremental proposals will always leave some groups unprotected; means-based coverage will always leave the poor vulnerable when the political will to meet social obligations flags.

One thing that is striking about the literature on children's health is the broad consensus that certain preventive and monitoring services—beginning prenatally and going through adolescence—are essential to protecting the health status of children. Many of the inequalities in health status between poor children, especially minorities, and richer children are traceable to lack of access to such services. At the same time, however, society is willing to lavish very expensive technologies on the "identified victims" that result from the absence of such preventive services. The point was brought home to me when I was taken on a tour of Houston a couple of years ago. I was shown the Jefferson Davis Memorial Hospital, which is located in a Chicano ghetto. It boasts, I was told, one of the most expensive neonatal intensive care units in the country—state of the art. Yet there was absolutely no prenatal maternal care outreach program in the area of the hospital. Moreover, after an infant surviving such treatment is released, it could end up uninsured again, since Texas has one of the most restrictive Medicaid eligibility standards of all states.

Is ours a prudently designed system? I will say something shortly about "big-ticket" items, like neonatal intensive care, and I am not saying that we should abandon such services. However, if I were designing a system I would have to live through at all stages of life, and if I wanted it to meet my needs early in life, I would prefer one that guaranteed access to prenatal maternal care to one that had no such services but offered neonatal intensive care instead. Prudence requires me to attend to the fact that the preventive program more cost-effectively protects and maintains normal functioning—and thus, my opportunity range—than neonatal intensive care. If resources were scarce, I would also prefer the prenatal maternal care program to various technologies that deliver less protection of opportunity range at higher cost at later stages of my life, e.g., aggressive chemotherapy for metastatic solid tumors. I am here not arguing rigorously, but only using the device of prudent lifespan allocation heuristically to help us see that certain allocations in our health system more effectively protect opportunity over the lifespan than others. On the theory that I am sketching, prudent allocations are the ones that are fair to different age groups, the ones that justice requires.

Similar points could be made about the recommendations of the Academy of Pediatrics and the Children's Defense Fund: the key difference in the health status of children will be made by relatively inexpensive, but labor-intensive services. Fetuses need their mothers to be adequately nourished, free of sexually transmitted diseases, and free of drugs. Children need immunizations, they need to be monitored for their physical and mental development, they need access to psychological and dental services, they need lead removed from their homes, and they need education and counseling as adolescents about sex and drugs. If we knew that we had to budget a lifetime fair share of health care so that it made our lives as a whole go as well as possible and so that it protected opportunity at each stage of life, we would not be stingy about these services. Indeed, we would trade access to them for access to many acute care services that are now provided as a matter of course and without any real concern about their costs. A very similar set of points could be made about access to long-term care for chronic diseases and disabilities: these services may be more important to protecting opportunity over the lifespan than lavishing certain technologies on the dying—elderly or young.

My point here is not that we should now ration big-ticket items, such as organ transplantation or neonatal intensive care, even if they would be less prudent to include in our system than services we now omit. I doubt that, in our wealthy system, the rationing of these big-ticket items is really necessary. In any case, we have hardly eliminated all services that are less cost-effective than these big-ticket ones—indeed, we are not even sure which ones they are. Nevertheless, under some conditions of scarcity—including the natural scarcity of organs—justice requires that we ration fairly, and so we must be clear about what principle should govern such rationing, including the dissemination of technologies whose opportunity costs are too high.

Institutional Obstacles to Just Rationing

I want to conclude by noting that it is not enough simply to have a good idea what an ideally just allocation of health care resource would be.[11] Suppose we conclude that protecting equality of opportunity over the lifespan implies that we should not disseminate as widely as we do certain big-ticket technologies, such as organ substitution technologies. Suppose, that is, that Oregon is right and that opportunity is better protected for the system as a whole by using our resources on other technologies.[12] The point is sometimes put by saying that the "opportunity costs" of some big-ticket technologies is too high. Knowing that a more just allocation would not include the widespread dissemination of these technologies will not stop their dissemination, for there are institutional obstacles to acting on these concerns about what justice requires.

Much of our health care decision making is not regionally or nationally coordinated. Different providers often directly compete with each other, which gives them some different, merely agent-relative rather than common, goals. Even where there is no direct competition, there is still no coordination or cooperation. When these providers pursue their goals—when they successfully pursue them—they may still produce outcomes that are collectively worse for them and us. It is important to see how this problem arises.

Suppose that we agree that, in an ideally just arrangement, the health care system would invest resources in services other than certain organ substitution technologies. On the other hand, if we forgo introducing this technology (supposing we are on a hospital board making such a decision), some other providers we compete with may do so anyway. Then those providers will be seen as the technologically most advanced hospitals and medical centers. They will attract physicians who seek the glamor and profit involved in offering such services; they will attract patients. They will be in a superior competitive situation to us. So if other providers introduce the new technology, we are better off if we do, too. On the other hand, if other providers refrain from introducing the technology although we add it, then we will enhance our competitive situation relative to them. Of course, each provider will reason this way. In the absence of political constraints that compel us to do otherwise, each will add the technology, and we will move farther from what we agreed would be a more just allocation of resources. Thus, when we all introduce the big-ticket items, the system works less well than it should, and our competitive situation is, in any case, no better than it would have been had we all refrained from disseminating the inappropriate technology.

It may seem that only bad motives could drive us to reason in this way, but this is not the case. Good motives can produce the same bad results. For example, we may be motivated to do the best we can for our patients; we may think that we have a special obligation to deliver all medically feasible benefits to our own patients. These motives will lead each provider to reason that it should introduce the new technology or risk, losing the opportunity to benefit its patients as much as it can.

Two important features of the US health care system create this problem. We allow health care institutions to operate competitively instead of requiring them to act cooperatively and collectively. Also, we do not force each institution to pay the price for introducing technologies with highly problematic opportunity costs. Together, these features create a context in which claims that justice requires spending resources in ways that do not involve certain technologies fail to obtain a grip on allocation decisions. This argument suggests that macro allocation decisions at the federal and state levels will have to establish specific priorities for technology dissemination and will have to provide the regulative and legislative muscle needed to produce cooperative and not merely competitive decision making.

Eliminating these features of our institutions will have to be an explicit priority for those who wish to make our system conform to acceptable ideals of justice for health care. Medical and other health care technologies have made it possible for us to make improvements in our well-being at all points in the lifespan. However, under real world conditions of scarcity, we cannot meet everyone's needs, and so we must figure out how it is fair to allocate limited resources. I have suggested a perspective from which to think about some of those questions. I have emphasized the importance of fair equality of opportunity in thinking about health care in general, and I have suggested that we use the device of thinking about prudent allocation over the lifespan as a way of answering questions about justice between age groups.

In pointing out that there are institutional obstacles that stand in the way of making our system more just, however, I do not intend to build skepticism about the relevance of thinking about justice to practical decisions. It is important to know that, even if people understand what would be just, institutional arrangements may make it impossible for them to make appropriate decisions. Even when we know we cannot fully achieve the ideal, it is still important to know just what stands in our way. What stands in our way in this case is not sacred or immutable, but only institutions that we have constructed and can alter.

The central thesis of this chapter has been that prudential allocation over the lifespan can guide us toward a policy that is fair to all age groups. My own account of justice for health care requires that the Prudential Lifespan Account be applied within a general framework that protects fair equality of opportunity. In the United States, two crucial features of the system stand in the way of it even approximating the requirements of justice. First, the financing of the system means that almost 40 million citizens are uninsured and many millions more are underinsured. Second, there is no mechanism for allocating resources in accordance with principles of fairness; indeed, the institutional obstacles we have just examined stand in the way. Together, these point to a policy conclusion. The most effective way to move the system into compliance with what justice requires would be to implement a national health insurance scheme that guarantees universal coverage and provides the means to carry out appropriate resource allocation measures, including rationing of beneficial services. I have not argued for this policy conclusion here, but do not see any way justice can be done without measures along these lines.

Notes and References

[1]This chapter is based on a talk delivered at the Conference on Justice Between Generations, University of Maryland at Baltimore, October 1988. I draw on material in N. Daniels (1988a) *Am I My Parents' Keeper? An Essay on*

Justice Between the Young and the Old. Oxford University Press, New York, NY, with permission.

[2]D. L. Frankfather, M. J. Smith, and F. G. Caro (1981) *Family Care of the Elderly.* Lexington Books, Lexington, MA, p. 1.

[3]I believe there are also problems with its retention of an employer-based system of private insurance.

[4]Oregon's position can be understood as follows: given that politically imposed resource limitations mean that some rationing by ability will take place, it is better to ration extrarenal organ transplants by ability to pay than prenatal maternal care services. *See* N. Daniels (1989) Comment: Ability to pay and access to transplantation. *Transplantation Proceedings* **21:3,** 3424–3425.

[5]*See* N. Daniels (1985) *Just Health Care.* Cambridge University Press, New York, NY, and (1988a) for a developed discussion of the approach to justice and health care sketched above.

[6]*See* J. Rawls (1971) *A Theory of Justice.* Harvard University Press, Cambridge, MA, pp. 75–90, 150–161, 175–183.

[7]This requires modifications of Rawls' equal opportunity principle, however. Cf Daniels, *Just Health Care,* pp. 39–55.

[8]The justification for this approach and many of its details are necessarily omitted here. *See* Daniels 1988a.

[9]For the details of the argument for this conclusion, *see* Daniels (1988a, Ch. 5). It would be very easy to misunderstand this argument if its more developed version is not examined carefully.

[10]D. Callahan (1987) *Setting Limits: Medical Goals in an Aging Society.* Simon and Schuster, New York, NY.

[11]The next several paragraphs draw on material in Daniels (1988b) Justice and the dissemination of big-ticket technologies, in *Organ Substitution Technology: Ethical, Legal, and Public Policy Issues.* Westview Press, Boulder, CO, pp. 211–220.

[12]*See* N. Daniels (1989) Comment: Ability to pay and access to transplantation. *Transplantation Proceedings* **21:3,** 3424–3425, for further discussion of Oregon and "ability to pay" as a criterion for organ transplantation.

Old Age and the Rationing of Scarce Health Care Resources

Mark H. Waymack

Introduction

There can be no denying that our society faces grave questions concerning the allocation of socially financed health care resources, both now and in the coming years. The increasing relative scarceness of these resources is the result of a variety of causes. Advanced technology is not only expensive in its own right, but it also increases health care costs by lengthening longevity. Furthermore, the promises of modern medical technology have served to increase the public's appetite for medical care, and as the costs have increased dramatically, private-pay patients and insurers have become unwilling to subsidize indigent care in the way that they used to. Hence, the government has been increasingly looked to as the source of resources for indigent care. Nevertheless, despite the vast expansion of public programs, the numbers of the uninsured have risen in recent years to around 39 million. Sections of our large cities, such as Washington, DC and Chicago, have infant mortality rates that rival or surpass many Third World countries, and AIDS has begun to exact a heavy toll, particularly on our urban public hospitals. To complicate matters even more, governmental resources have become increasingly constrained as the federal government attempts to come to grips with a federal budget that now devotes nearly 20% of its outlays to deficit financing. To cap all of this, the numbers of the elderly, a segment of the population that uses the highest per capita amount of health care resources and that is universally covered by governmentally funded Medicare, are increasing at an enormous rate.

Even if one wishes to discount the most pessimistic of the doomsayers, it cannot be denied that resources are scarce. Rationing is already being done, largely in terms of ability to pay. Those eager to promote a more thoughtful, equitable, and moral scheme of rationing have begun to offer various proposals. One that has been frequently discussed in the last two years is the notion that old age should serve as a rationale for denying certain kinds of socially financed health care, particularly life-extending care, for the elderly.[1]

In this chapter, I shall aim to do two things. First, I shall review the philosophical reasoning in the views of four well known voices in this debate: former governor Richard Lamm, Norman Daniels, Daniel Callahan, and Larry Churchill. Second, I wish to argue that, although I do believe that a morally just scheme of rationing could (and should) include certain constraints on what kind of care is extended to the elderly, such a scheme will elude our understanding and acceptance unless and until we, as a community, have a much richer and more reflective conception of the nature and meaning of old age.

Richard Lamm's Utilitarianism

Lamm made headlines when, as governor of Colorado, he publicly stated that the debilitated, chronically ill, very elderly had a moral duty to forgo further health care and to accept their deaths.[2] As might have been expected, though regretted, the furor that ensued paid little attention to Lamm's reasoning.

Reviewing Lamm's position, it becomes apparent that Utilitarian considerations are his foundation. For instance, when Medicare was first enacted, the elderly suffered disproportionately from the travails of poverty, but, Lamm[3] argues, times have changed, and it is children who now live in poverty with far greater frequency than any other segment of the population, including the elderly. Lamm's utilitarianism becomes even more evident when he points out that, because children have far more years ahead of them than the elderly in which to suffer or to flourish and be happy, they deserve a greater share of our resources. Thus:

> It is sad that we now have a system that allows the elderly to consume far more medical resources than we give to children. It is not only fair but desirable to have a different level of care for a 10-year-old than for someone who is 100. Should not public policy recognize that some people have far more statistical years ahead of them than others? I feel it is morally repugnant to use $100,000 or more of our kids' limited resources, as I'm on my way out the door.[4]

Clearly, the working ethical principle here is that our resources should be distributed in such a way as to maximize the general happiness, not just now, but in the long run. As Norman Daniels describes this position:

> [S]ome urge that it is more cost-beneficial or cost-effective to use certain resources to save the young rather than the old because of the difference in "quality adjusted life years" saved. ...[T]here is a better return on the social investment of resources if the young get treated and the old do not.[5]

Even a committed Utilitarian, however, would pause at Lamm's conclusions. John Stuart Mill, of course, was the author not only of *Utilitarianism,* but also of *On Liberty.* What we all must recognize, including the Utilitarian, is that any society that does not respect the individuality of its members, even when such respect may have the *appearance* of running counter to the general happiness, will not be a flourishing, happy, and humanly rich society. Also, any society that withholds health care from the elderly on the grounds that the "return" on such an investment will be low because of limited remaining lifespan runs the danger of being just this sort of society—unhealthy, unhappy, and divided against itself.

The challenge that Lamm's observations force on us, therefore, is how to allocate our resources in such a way as to provide for the long-term health and happiness of society as a whole while not abandoning the elderly as nothing more than unprofitable investments not deserving of further consideration and concern. It is with these concerns in mind that we turn to the Rawlsian liberalism of Daniels.

Daniels' Rawlsian Liberalism

In contrast to the Utilitarian approach, Daniels (borrowing key features from Rawls' theory of justice[6]) offers what he terms a prudential theory of resource allocation. If we each had unlimited resources available to us, such a budgeting plan would not be necessary, but given the reality of limited (though certainly not meager) resources, it becomes only sensible to plan for the long run rather than to squander all of our resources at any one lifestage. According to this prudential theory, the individual should therefore consider how he or she would allocate what limited resources he or she has available over the different stages of life as a whole. We are each to consider the possible course of our lives as a whole, with impartiality between different stages of our lives, and develop a rational plan for budgeting our resources between our childhood, youth, middle age, early old age, and very advanced age.

If we are to be able to pursue whatever ends we may have at any particular stage in our lives, it will be requisite that we be not only alive, but able to function at a tolerably normal level for that age. Thus, Daniels thinks that, being sensible individuals, we will each choose to budget our resources in such as way as to best ensure our abilities to function at the ordinary level of functioning for each stage of our lives, a level that Daniels terms, *normal species functioning*. With respect to allocating different amounts of health care resources to different age brackets, the rational individual will seek a budgetary compromise between, on the one hand, throwing *all* one's resources into the earlier stages of life to increase the odds of making it to old age and, on the other, avoiding the misfortune of living into old age with no resources left to lessen one's suffering. According to Daniels, therefore, under conditions of *severely* limited resources—a situation where allocation of substantial resources to our later years would *significantly* increase our likelihood of not living past our youth—an age-based rationing scheme could be rationally chosen by the individual if it was seen as the best possible means for ensuring us a relatively normal lifespan.

Having arrived at the conclusion that we would allocate our health care budget so as to best ensure our opportunities for normal species functioning at each stage of our lives, Daniels believes that the impartiality and fairness of justice demand that we extend that plan as social policy for society as a whole.

One point bears emphasis. It is not Daniels' position that scarce health care resources should be used inefficiently; rather, his objection to the kind of position put forward by Lamm is that such a crassly Utilitarian procedure ignores the life plans chosen by the individuals themselves. As Daniels puts it:

> It is an important feature of the prudential argument I offer, however, that it involves no judgments by one person about the value or worth of another's life. Instead, it involves persons making judgments *for themselves* about benefits *to themselves* at different stages of their lives.[7]

There are two difficulties, however, that merit discussion. The first concerns the extension of individual choices to social policy, a move that Daniels must make, but a move that is complicated by his commitment to liberalism. The second concerns our construction of a rational life plan that is to serve as the basis for our budgetary allocations over an entire lifespan. Each of these, I believe, poses quite troublesome philosophic difficulties.

The moral attraction of the frameworks offered by Rawls and Daniels is that, though resources may be allocated in restricted ways, they are allocated in ways that *each of us as an individual* would choose from a viewpoint of strict impartiality. Hence, the liberal value of *individual* autonomy is preserved, while *social* policy can be created and enforced. Unfortunately, this happy coincidence of agreement rests on the assumption that, as individuals, we will make generally the same decisions under conditions of impartiality. However, some individuals may place quite different values on health, life, and its different stages than others. Hence, instead of each individual's budgetary decisions being in harmony with those of everyone else, there may well be irreconcilable disagreement. What sort of allocation policy is just will then depend on what particular individual we ask, even when the individuals are thought of as being strictly impartial (as in Rawls' "Original Position"). As Engelhardt argues against Rawls (and the same point will hold here against Daniels):

> [The answer] will depend on how they rank particular harms and benefits. In short, Rawls presupposes the legitimacy of a particular moral sense. He is best understood as having engaged in a much more limited, but still important, goal of providing a rational reconstruction of the moral world of a liberal member of the Cambridge, Massachusetts, community.[8]

Our second objection concerns not interpersonal disagreement, but the difficulty of developing an individual rational life plan in the first place. In *Am I My Parents' Keeper,* Daniels devotes an appendix to replies to objections raised by Derek Parfit concerning our willingness to be impartial between our different "selves" at the different stages of our lives. What concerns me here is even logically prior to Parfit's concern. *Is it possible for us to imagine, in any helpfully full and realistic sense, what our lives will be like in our old age and what sorts of values we will hold at that stage in our lives?*

The difficulty posed here is twofold. It is not just one of imagining the self at an age that one has not yet experienced (the Shirley MacLaines of the world aside); it is also the daunting task of defining and integrating the values of that age with the values of the rest of one's life. This is not a power of the imagination that one comes by easily. It requires not only a vivid and enlivened imagination; it also requires a well equipped vocabulary and a practiced repertoire of biographies, so to speak, in order to understand one's values in old age and how to integrate them into the complex autobiographical novel of one's individual life as a whole.

I suggest that the cultural and psychological diversity of our society makes it unlikely that even behind Rawls' "veil of ignorance" we will reach the kind of agreement that Daniels envisages concerning the allocation of health care resources. However, even worse than that, I fear that a general poverty of imagination and understanding in our society concerning old age will prevent the vast majority of individuals from even being able to construct coherently the kind of rational life-plan health care budget that Daniels requires of us.

Callahan's Communitarianism

Callahan's position is that:

1. The government has a duty to help people live out a "normal lifespan," but not to live beyond the normal lifespan through medical care.
2. The government should devote resources to developing and employing medical technology and care that assist people in living a normal lifespan.
3. The government should provide only palliative care for those persons who have already lived out a normal lifespan.[9]

As we shall see, Callahan uses two quite distinct, though intertwined, arguments to defend this position. The first line of argument is that there is a "normal lifespan." It is morally obligatory, out of a sense of equality of opportunity, to assist persons to live out their normal lifespans, but it is neither obligatory nor helpful to assist persons to live beyond the normal lifespan. This is what I shall call the "Normal Lifespan Argument." The second line of argument is that, in allocating social resources, *scarce* resources must be allocated *efficiently* if society is to continue to thrive as an ongoing community, a community that exists through generations.

Though Callahan does intertwine these arguments, they have quite different motivations, justifications, and implications. It will be useful, therefore, to consider them separately before returning to an overall assessment of Callahan's project.

The Normal Lifespan Argument

As he notes, Callahan borrows his normal lifespan argument from Daniels.[10] As Daniels presents that argument, the initial premise is that we have a social obligation to provide equality of opportunity to the various members of society. Equality of opportunity is normally thought of in terms of educational or job opportunities, but as Daniels points out, ill health can be debilitating and constraining in its own right. The person suffering from congestive heart failure or surgically correctable cataracts, but who does not have access to proper medical care, has his or her opportunities in life severely constrained in a way that is not *normal.* Furthermore, this is often the result of the genetic or environmental lottery, a lottery that seems to offer no *moral* justification of why some people are afflicted with medical suffering, but others are not. Hence, Daniels concludes that, if we are to pay anything more than empty lip service to equality of opportunity, then we must acknowledge and act on a moral–social obligation to ensure as far as we are able that ill health does not become a barrier to equality of opportunity.

Callahan sees this argument as morally requiring the social guarantee of therapeutic care (as well as chronic and palliative care) for the young and middle aged of society. However (in a way that does not make Daniels entirely happy), Callahan concludes that this moral–social obligation does not extend to therapeutic health care for those of advanced age. This is because society's obligation is only to ensure the *normal* range of opportunity, and since it is anything but normal to live an indefinitely extended life, society has no obligation to provide the means necessary for that project.

There is a striking, superficial similarity between the arguments of Daniels and Callahan, but as we shall see, their roots and implications are really quite different. Unlike Daniels' argument, the normal lifespan argument in Callahan's work is motivated largely by a concern that modern society's and modern medicine's attempts to deny Callahan's sense of a normal lifespan leave the elderly themselves worse off than would a social attitude of understanding and acceptance of the inevitability and appropriateness of death in old age. In itself, this goes beyond typical arguments in terms of distributive justice. It is really an argument about what constitutes "the good life." It is about what is *really* in the best interests of the elderly themselves. It will be helpful to unpack this argument a bit here.

First, Callahan offers what we may term a phenomenological account of the meaning of old age. He vehemently rejects the view that old age should be understood as a time for detachment from obligations to others, and for devotion to recreation and self-indulgence. Such a view, Callahan argues, perpetrates a destructive myth; it creates a vision of old age that fosters disenchantment, alienation (in the classical sense), and a diminished sense of self-worth. Instead, Callahan urges, for old age to have a deeper, more valuable meaning for both society and the elderly themselves, we should understand old age as having a continuing social role. In particular, he suggests that an essential feature of the role of society's elderly is the communication of society's history and values to younger generations. Also, and of central importance to our topic of discussion, he feels that an appropriate sense of the passing of generations, one to another, is a key to understanding our positions as members of a generation in a society that has its roots in the past and that extends into the future. Thus, an essential part of the meaning of old age (and indeed, of life itself) is that at some point it becomes *appropriate* to accept death. Though we may be saddened, for example, by the death of a 90-year-old parent and grandparent from a heart attack, we do not regard such a death with the same sense of tragedy or injustice as the death by heart attack of a 40-year-old father or the death of a 7-year-old child from leukemia. The death of the 90-year-old may be seen as a "passing of the torch" from one runner to the next. The death of the 40-year-old, however, is like that of one who stumbles in the middle of the race course, whereas the death of the 7-year-old will be seen as that of someone who *never* got to run his or her leg of the race at all.

From this phenomenological analysis of the meaning of old age (and I certainly have not done full justice to Callahan's analysis), Callahan concludes that the elderly themselves are better off if they are engaged in this social role (as described by Callahan), and this engagement includes the acceptance of death as appropriate and not to be avoided in advanced old age (perhaps beyond 85 years). Thus, it is better for the elderly if we do not offer them any therapeutic medical care beyond this threshold.

Note how clearly this position is different from that of Daniels. In Daniels' vision, a just society, if it possessed abundant resources, might legitimately choose to engage in research in and treatment with *therapeutic* geriatric medicine, therapy aimed not only at ameliorating the suffering of the elderly, but at prolonging their lives. From Callahan's point of view, however, such endeavors would be immoral, because they would foster misguided hopes, create unrealistic and unsatisfiable expectations, and undermine any healthy sense of the meaning of old age, replacing it only with a distressing and debilitating sense of *anomie*. It is the very nature and meaning of old age, in this argument, that morally justifies refraining from research into the provision of life-extending technology for the elderly.

If, for the sake of argument, we were to accept Callahan's phenomenological analysis of the nature and meaning of old age, it would still not be the case that the propriety of refraining from research in geriatric therapy and refraining from offering life-extending procedures to those of advanced age would follow directly from the phenomenological analysis. The missing (or at least very obscure) piece of the argument is that society does not have a moral obligation to assist its individual members in their pursuit of their *own* conception of the good life when their conception differs from what we take to be the more correct or appropriate conception. In other words, even if Callahan's analysis of the meaning of old age is in some sense right, does this absolve society of any moral obligation to help some of its members live out a different view, for example, a view that holds it is good to prolong life as long as possible?

Daniels, for example, would reject such a position. Holding to a very "thin" conception of the good, Daniels' liberal position places a moral obligation on society to help foster the conditions that allow individuals to pursue their *own* personal, "thick" conceptions of the good, conceptions that might include indefinitely extended longevity.

What Callahan needs here, and what *Setting Limits* lacks, is a well argued moral–political philosophy that would justify the restricting of social benefits according to certain thick conceptions of the good. Our conclusion would have to be that society ought not to subsidize the pursuit of alternative conceptions of what constitutes the good life, including a good old age. Such an argument, though not necessarily unsuccessful, would have to face squarely a long American tradition of political and moral liberalism, a political philosophy that has endured despite then-candidate Bush's demagogic castigation of the *word* "liberal" in the 1988 presidential campaign.

The Societal Efficiency Argument

Though he appears hesitant to use it as an argument, Callahan repeatedly shows concern for the burdensome and rapidly increasing costs of health care for the elderly in relation to other important needs of our society. What motivates this efficiency argument is the conception of society not as a static collection of self-interested individuals, but rather as an ongoing project continuing *through* generations. One of the obvious, and morally quite appropriate, goals for society (as understood in these terms) is that *society be able to continue to exist and to flourish.* Once we see society as an endeavor extending from the past and into the future, the moral appropriateness of the transfer of wealth between the generations can become comprehensible. For if society is to continue, and not just at a subsistence level but at a flourishing level, then if its resources are scarce, they must be allocated in a manner that is most efficient for *sustaining* a flourishing society. Beggaring one's children for the sake of one's grandparents would be morally backwards.

This is clearly an argument cast in terms of distributive justice. Furthermore, it provides reasons for restricting expensive therapeutic care from the elderly, even when some of those elderly might desire such aggressive treatment. The justification, of course, would be that scarce resources could be put to much more effective use from a long-term perspective (e.g., prenatal care, childhood immunizations, and even nonhealth care goods or service, such as quality education for the young, might take precedence over life-extending therapy for those of greatly advanced age).

It is easy to see how this argument could lend itself to discrimination on the basis of age, though we should be quick to note that discrimination *per se* is not necessarily morally repugnant. (Surely, it is morally acceptable to discriminate between fine and bad wine, between good and bad literature, between good and bad actors, and so on.) For one thing, exotic and expensive interventionist therapies usually do not succeed with the elderly as well as with younger people. This, for example, has become increasingly clear with respect to the efficacy of cardiopulmonary resuscitation. More importantly, though, the future of society lies far more clearly with the young than with the old; it is the young who will grow to fill the shoes of industry, leadership, and parenthood that an older generation now fills. Hence, to indulge the presently very old aged at the expense of significant harm or deprivation to the young would be *unjustly* detrimental to several generations to come.

We should note that this efficiency argument does not *necessarily* work itself exclusively along age brackets. Nor does it necessarily guarantee the inclusion of palliative care and the exclusion of therapeutic care for the very aged. It may well entail that expensive care for the elderly (therapeutic or palliative) would need to be withheld for the long-run benefit of society as a whole. However, it may also be the case that some relatively inexpensive yet therapeutic care *should* be provided to the elderly, even if doing so would require denying extremely expensive care that is either of marginal benefit or benefits only a very few persons in the younger age brackets. Doing so may advance the overall level of flourishing in society, both now and in the long run.

The following example may help illustrate the point: Consider two different patients. On the one hand, there is a 50-year-old woman diagnosed as having a rare clotting disorder that will be fatal if untreated, and that will be chronic but manageable if treated. The only effective course of treatment, however, is the very expensive monoclonal manufactured human clotting factor VIII. Because the patient lacks both insurance and personal funds, this will cost society over $1.7 million. On the other hand, there is an 87-year-old woman who has a fractured hip. She is in otherwise very good health and, in her physician's judgment, could benefit substantially by an artificial hip replacement as opposed to less aggressive therapy.[11] Total care for her would be only in the tens of thousands of dollars. Callahan himself, in *Setting Limits,* acknowledges that a sense of justice would lead us to treat this 87-year-old woman, but though his admission may strike us as a morally correct intuition, it is not at all clear how such an exception can be made consistent with his professed conclusions. It would appear then that the societal efficiency argument does not lead directly to Callahan's stated theses.

One possible response would be to argue that, whereas "absolute" justice might prefer the 87-year-old in the above two cases, *public policy* works best when its directives are relatively clear, easily understood, and capable of application with little or no vagueness or ambiguity. This is because public policy, if it is to have substantial public support, must be easily comprehensible by the public and must be of a form that is straightforward and easy to apply, such that there can be few or no questions about procedural fairness. With those requirements in mind, using the line demarcating old age from the rest of the population as a workable distinction to deny any social obligation for therapeutic care may capture most of the determinations made under the efficiency argument.

Such an argument, however, seems quite weak. For one can imagine other schemes, such as one that excludes certain therapies that are relatively ineffective or excessively expensive, that capture our intuitions about justice in these matters more closely.

We may understand now why Callahan closely intertwines what I have termed his normal lifespan argument with the argument from societal efficiency. Neither, by itself, directly yields the kind of results that Callahan thinks we should accept. In a way, they are mutually dependent—the efficiency argument shows why rationing cannot be avoided, and the natural lifespan argument provides a far more sophisticated rationale for rationing against the very elderly and in favor of the young than does a naive sort of utilitarianism.

What remains at the heart of Callahan's argument, therefore, is his conception of the nature and meaning of old age. Interestingly, however, the vast bulk of the responses to Callahan, particularly the vehemently antagonistic public outcries, have paid very little attention to his phenomenological analysis of old age and its meaning. Completely overlooking (perhaps willfully ignoring) Callahan's attempt to articulate reasons why *age itself* might make a morally relevant difference, the popular response has been in terms of rights and age discrimination.[12] Why is this the case? Once again, my suggestion is that as a society we do not now possess a rich enough conception of old age and its meaning and value to engage profitably in such a debate. Without such conceptions at hand, and without a vocabulary in which even to frame the debate, we (society at large) are unable really to see, much less digest and react to, Callahan's phenomenology.

Churchill's Justice-by-Sympathy

Churchill's *Rationing Health Care in America*[13] shares Callahan's rejection of liberalist individualism. It also rejects the conception of society and social justice of the naive Utilitarian. Like Callahan, Churchill urges upon us a communitarian sense of justice. Though Churchill does not discuss the question of age-based rationing as explicitly as Callahan, his exposition and defense of a communitarian sense of justice offer us some constructive clues as to how to lead progressively forward a discussion of Callahan's communitarian-linked conception of the nature, meaning, and value of old age.

Sympathy

Hearkening back to the Scottish Enlightenment, Churchill presents an understanding of morality and justice drawn particularly from Adam Smith's discussion of *sympathy* as the means by which we are able to engage in morality. Humans are, by nature, *social* creatures and morality is, in its essence, a social phenomenon. It is by *feeling-with* other persons (*sym-pathy*) that we are able to take an interest in the welfare of others, to comprehend what their interests might be, and to adjudicate when the interests of someone else conflict either with our self-interest or the interest of others. As Smith explains in *The Theory of Moral Sentiments:*

> As we have no immediate experience of what other men feel, we can form no idea of the manner in which they are affected, but by conceiving what we ourselves should feel in the like situation. Though our brother is upon the rack, as long as we are at our ease, our senses will never inform us of what he suffers. They never did, and never can, carry us beyond our own person, and it is by the imagination only that we can form any conception of what are his sensations. ...By the imagination we place ourselves in his situation, we conceive ourselves enduring all the same torments, we enter as it were into his body, and

> become in some measure the same person with him, and thence form some idea of his sensations, and even feel something which, though weaker in degree, is not altogether unlike them.[14]

In terms of moral judgment, Smith argues that we come to approve or disapprove of other persons by sympathetically imagining ourselves in their position and considering whether we would share their sentiments and motives or not. Furthermore, since the goal of morality is to guide our motives and behavior in ways that are agreeable and useful not only for ourselves, and not only for those close to us, but for society *as a whole,* we should, in our moral judgments, attempt to sympathize (via our imaginations) both as accurately as possible and over as wide a collection of persons as possible. This process both requires and perpetuates a sense of *community.* As Churchill puts it, "we have seen that a way out of this problem is a renewed sense of community, grounded in our awareness of our social character as human beings and our innate affinity for social life."[15]

Sympathy and Age-Based Rationing

In terms of determining a morally rationing scheme involving old age as a factor, we must endeavor to sympathize (in Adam Smith's special sense of that term) with the elderly, but also with all the other age brackets of society as well. When we think in this fashion, Churchill argues that we will recognize that unlimited care for the elderly, particularly life-extending care "at the margins," is not morally appropriate. He notes that certain infectious diseases and medical conditions require of the physician reports to public authorities for the welfare of the public health. Certain venereal diseases are reportable, despite the harm that may come to the individual in question. Significant risk for cardiovascular accidents in airline pilots are reportable, as should be similar conditions for school bus drivers. In all such cases, the constraints on a right to privacy are justifiable in terms of a greater common good. Pointing to this as a widely accepted moral precedent, Churchill proceeds to argue:

> Christian doctrines of stewardship prohibit the extension of one's own life at a great cost to the neighbor. This is not fatalism but a simple matter of proportion. Most patients would not bankrupt their family and deny their children a fair start in life by striving for a last, expensive extension of their own lives. An imaginative sympathy shows that neither should we extend our lives at the margins if by so doing we deprive nameless and faceless others a decent provision of care. And such a gesture should not appear to us a sacrifice, but as the ordinary virtue entailed by a just, social conscience.[16]

Thus, Churchill concludes, "If the next generation is to flower and flourish we must practice the wisdom of giving ground when our time comes."[17]

Some Difficulties

Churchill himself writes about renewing a sense of community. His argument, then, is a plea to restore, rejoin, and renew our sympathetic sense of community. That this needs to be a project of renewal is painfully evident. As I write this, elderly persons in New York are in court fighting for their "right" to live in housing communities that exclude any and all children. Apparently, these elderly see their social obligation to tolerate children as a burden that they have finished with. Yet, at the same time, the idea that extended Medicare benefits (which the elderly wanted) should have been funded by a modest tax on the elderly themselves created an acrimonious hue and cry that has led to swift repeal. Also, the popular media, weighing in on the side of the younger, have represented the elderly as "greedy geezers," the most affluent age bracket of current society, who are somehow to be blamed for a younger generation's diminished ability to own their own homes.

Churchill's justice, his rapprochement, depends on our ability *and* willingness to sympathize imaginatively with a broad spectrum of other persons. However, are his hopes in this vein realistic? It is to that question that we must now turn.

Toward a Shared and Rich Conception of Old Age

My own conviction is that the lines of argument taken by Churchill and, more specifically, by Callahan are for the most part morally correct. However, I believe that it is premature to think that the argument can be made convincing to a wider public. As a step toward a conclusion, I shall offer some thoughts on the obstacles that we face in reaching a sense of justice in this matter, and some suggestions on how we might help to move the public forum forward in a constructive way.

Some Obstacles

A point that we have seen emerge repeatedly is that our general inability to form a common notion of the meaning and value of old age prevents the public debate from progressing. There are two (related) ways in which we are unable to do this. First, the social distancing between the generations has diminished the ability of the younger to imagine the predicaments, hopes, and fears of the elderly. Likewise, it has diminished the ability of the elderly to imagine vividly the problems, needs, and constraints that face their progeny. The disappearance of the extended family and the isolation of age brackets among their own kind are both causes and symptoms of this fragmentation of the community.

Second, the rapid changes in life-extending medical technology, the historically unprecedented aging of our society, and the youth-oriented attitude of our culture for the last several decades have all come together to make it exceedingly difficult for us—young, middle aged, and elderly themselves—to have a conception of what a *good* old age would be. Physicians and ethicists sometimes imagine that, if one can learn the earlier "biography" of the elderly patient, then one can interpolate therefrom to a fairly clear notion of how the story should proceed from the present moment. This, of course, can only succeed if one already has a general idea of what good closing chapters might look like. However, our youth-oriented culture thought very little about what it meant to be old, and what remnants of our understanding and conception of old age we retained have been of little applicability because of the dramatic newness of the situations that our elderly and our society have faced. Elderly patients and their families are often baffled when physicians ask them for guidance concerning medical treatment in the last few weeks, days, or even hours of the lives of the elderly. Unlike the days, decades ago, when the death of an elderly family member occurred at home in fairly familiar and predictable ways, advanced medical technology has both institutionalized death (usually in the hospital or nursing home) and altered its course, such that the patient and family are often left puzzled, incomprehending, and speechless concerning how the biographical novel should appropriately end. Also, because of our social alienation from this encounter with medically managed death, those families that have struggled through the mystifying ordeal are reluctant or unable to share meaningfully their experiences with the other members of their community.

Churchill is right, then, that we need a "re-visioning," but how is the discussion to proceed? How many novels, how many films, and how many popular television shows present us with realistic, coherent, and meaningful examples and interpretations of goodness, meaning, *and limits* in old age? Regrettably few.

Here, one must enthusiastically applaud Callahan, for although the popular mind has thus far chosen to react with vehement indignation (and as one who has tried to explain and defend elements of Callahan's argument before an audience that included a number of Grey Panthers, I am all too familiar with such acrimony), the topic has been brought into the open. As more "stories" circulate—examining, probing, and constructing a deeper social meaning of old age within our modern society (with its burgeoning numbers of elderly and its life-extending medical capabilities)—I am confident that the wisdom of Callahan's proposal will become more evident.

Unfortunately, Callahan may be correct in suggesting that such a philosophical, social project may take several decades to achieve. I say unfortunately, for in the meantime, our children, our society, and our elderly themselves will continue to suffer alienation, deprivation, and *anomie.*

Notes and References

[1]The catalysts for discussion have chiefly been D. Callahan's (1987) *Setting Limits.* Simon and Schuster, New York, NY, and N. Daniels' (1988) *Am I My Parent's Keeper?* Oxford University Press, New York, NY. In October 1988, the Maryland Humanities Council sponsored a conference on the topic, and the April 1989 meetings of the Society for Health and Human Values were devoted to the issue.

The issue has made its way into medical journals, such as in L. Churchill's (1988) Should we ration health care by age. *Journal of the American Geriatric Society* **36,** 614–617. It has been discussed by H. Moody (1988) in Natural limits to health care. *Medical Humanities Review* **2,** 31–38, and it has been in philosophical journals, as in D. Brock's (1989) Justice, health care, and the elderly. *Philosophy and Public Affairs* **18,** 297–312.

[2]Although Lamm's views have been frequently heard through the audio-media, they have made it into print less often. I shall base most of my remarks on an article by Lamm (1989) Nine commandments for an aging society. *The Aging Connection* **10,** 8,9.

[3]Lamm, *The Aging Connection* **10,** 9.

[4]Lamm, *The Aging Connection* **10,** 10.

[5]Daniels, *Am I My Parents' Keeper?* p. 94.

[6]*See* J. Rawls (1971) *A Theory of Justice.* Harvard University Press, Cambridge, MA.

[7]Daniels, *Am I My Parents' Keeper?* p. 94.

[8]H. T. Engelhardt (1986) *The Foundations of Bioethics.* Oxford University Press, New York, NY, p. 36. For the same sort of point, *see also* B. Williams (1985) *Ethics and the Limits of Philosophy.* Harvard University Press, Cambridge, MA, pp. 77–80. A similar point was argued by N. Jecker against Daniels at the conference, *Justice Between the Generations,* October 1988, at the University of Maryland Baltimore County.

[9]Callahan, *Setting Limits,* pp. 137,138.

[10]Callahan, *Setting Limits,* p. 137.

[11]This example also illustrates the difficulty Callahan faces in distinguishing clearly between therapeutic and palliative treatment.

[12]Few people stop to note that our Federal government already engages in age-based discrimination. Of the approx 39 million persons in this country who have no health insurance or are not eligible for government insurance, none of

them are over 65 years of age. Medicare is a federally financed program designed for the benefit of the aged.

[13]L. Churchill (1988) *Rationing Health Care in America.* Notre Dame University Press, South Bend IN.

[14]A. Smith (1976) *Theory of Moral Sentiments* (A. Macfie and D. Raphael, eds.) Oxford University Press, Oxford, UK, p. 9.

[15]L. Churchill, *Rationing Health Care in America,* pp. 104,105.

[16]L. Churchill, *Rationing Health Care in America,* p. 112.

[17]L. Churchill, *Rationing Health Care in America,* p. 112.

Appeals to Nature in Theories of Age-Group Justice

Nancy S. Jecker

> An elderly dame, too, dwells in my neighborhood, invisible to most persons, in whose odorous herb garden I love to stroll sometimes, gathering simples and listening to her fables; for she has a genius of unequalled fertility, and her memory runs back farther than mythology, and she can tell me the original of every fable, and on what fact every one is founded, for the incidents occurred when she was young. A ruddy and lusty old dame, who delights in all weathers and season, and is likely to outlive all her children yet.[1]

Appeals to nature are heard with greater frequency in contemporary bioethics literature. Often, such appeals ground arguments defending limits to health care for older age groups. This chapter examines how the language of nature enters moral arguments and raises problems for theories of age-group justice.

Normal Species Functioning

The idea of nature evokes no single image or meaning, but instead is common ground to a considerable range of ideas. On the whole, different senses of nature elicit positive images. High,[2]

Aging and Ethics Ed.: N. Jecker

for example, enumerates seven different senses of natural death in ordinary language. Each either attributes positive value to nature directly or does so indirectly, by representing nature as orderly or by suggesting that what is natural is as it should be. Thus, a natural death is thought to result from biological processes of age or disease rather than from accidents or violence, or a natural death suggests that dying is natural for biological beings. Other conceptions view a natural death as inherent or spontaneous, rather than contrived or prolonged; easy, peaceful, and comfortable; expected, not startling or surprising; a necessary destiny or what is universal to a species, not contingent; and expressing the law-abiding harmony of nature.

That nature is considered in a positive light in ordinary language motivates its skillful use in philosophical argument. Philosophers who appeal to concepts or images of nature can rely on its diverse favorable connotations to lend support to their arguments. For example, adjoining the word "natural" to "death" already suggests a good death. Likewise, the idea of a natural lifespan already prompts the thoughts that such a span of life is fitting and right, and that living beyond it is suspect. Similarly, calling certain functioning "species typical" carries with it the idea that improving functioning beyond this point is extraordinary, not obligatory. Thus, although the history of ethics teaches that *ought* statements cannot be derived from *is* statements,[3,4] this does little to deter such appeals. For, despite ourselves, we tend to think that what is ought to be. The upshot of this is that those attuned to the added force that appeals to nature give often refer to *is* statements in order to lend support to *ought* statements.

A good illustration of this is found in Norman Daniels' recent work on distributive justice in health care.[5] Daniels begins by introducing several basic concepts that ground a theory about justice between age groups. He first defines diseases as "deviations from the natural functional organization of a typical member of a species" and health as "the absence of disease."[6] These definitions suggest to Daniels a reason why meeting health care needs is important: the presence of health or disease dramatically influences individuals' normal species functioning, which in turn affects the range of opportunities normally open to people. Daniels calls the opportunities that are normal at each stage of life the "age-relative normal opportunity range."[7]

For Daniels, the idea of age-relative normal opportunity helps define the scope and limits of just health care for different age groups. In particular, Daniels argues that, if each of us were to deliberate under a veil of ignorance about how we would like scarce medical resources distributed over our lifespans, we would consider it more important to devote resources to diseases or disabilities that diminish our age-relative normal opportunities. The conclusion Daniels draws is that a just health care system is required to restore or compensate only for those conditions that impair opportunities that are normal at each age. In addition, age rationing is justified whenever it can accomplish an increase in a person's chances of living a longer-than-normal lifespan only at the cost of reducing that person's chances of ever reaching a normal lifespan.

Notice how appeals to nature play a pivotal role in Daniels' argument. At the start, the effort to construct a viable ethic for distributing scarce resources is aided by the belief that certain levels of human functioning are normal. The idea of normal functioning is itself grounded in biology or nature. First, normal functioning refers to species-typical functioning rather than functioning that is statistically average. Second, a species is a biological order that exists in nature. Finally, members of a species function according to laws of nature or patterns that are universal for their species. The idea of normal opportunity range also involves an appeal to nature and biology, since what falls within this range is largely determined by normal species functioning together with general facts about one's society.

An appeal to nature and biology adds considerable force to Daniels' argument on several scores. First, if what is species typical or biologically normal is considered good, then the suggestion to restore or compensate for deviations from this standard appears correct. Furthermore, requests for more than this appear suspect. Finally, because aging is understood as a physical process and biological destiny, the restrictions placed on health care for older age groups appear legitimate as well. After all, if it is natural or species typical to experience diminished functional capacities and a smaller range of opportunities, then the aged cannot reasonably expect to function at the level younger age groups do, or have the same range of opportunities available to them.

In response to the above remarks, Daniels might acknowledge that normative aspects color our concepts of health and disease. To the extent that they do, he might claim, we are not referring strictly to health and disease, but to wellness and illness. On this reading, the biomedical model of disease allows us "to make normative judgements *about* diseases. ...[However,] these normative judgements yield the normative notion of *illness,* not the theoretically more basic notion of disease; [moreover,] this distinction admittedly departs from looser ordinary usage."[8] According to this perspective, the concept of disease is not itself marked by normative assessment. Whenever normative assessment enters our thinking, we are no longer thinking about disease in the strict sense.

Does this response succeed? In order for it to do so, we must be willing to say that Daniels' definition of disease in terms of "the natural functional organization of a typical member of a species" constitutes a purely descriptive account. Clearly, though, this is not the case. As the preceding discussion shows, the appeal to nature that is part and parcel of this account is hardly an appeal to "hard fact." It summons rich normative meanings: nature manifests good, represents what is necessary, commands obedience, and so on. Thus, Daniels can hardly define disease in terms of nature while simultaneously keeping the idea of disease descriptive and untainted by normative judgment. Given that the idea of nature itself is not immune from normative notions, the definition of disease in terms of nature cannot be lexically prior to normative judgments either.

Natural Lifespan

Another example of appealing to nature to generate normative results is found in the work of Daniel Callahan.[9] Callahan's argument for restricting health care for older age groups begins by setting up a sharp contrast between images of medicine and technology run amuck, on the one hand, and images of benevolent nature and a fitting natural order, on the other. He first describes a mode of thinking marked by inveterate optimism and a faith in aggressive scientific research, and points out how this mode of thought has prompted a change in our criteria for premature death. Whereas previously death was seen as "a common occurrence at all stages of life and the attainment of old age the good fortune of a small percentage only," today, age 65 is the dividing line for premature mortality and normal lifespan.[10]

Our heightened expectations trouble Callahan because they suggest to him that our sense of what constitutes a reasonable length of life is held hostage by modern medicine: "if concepts such as 'old,' 'aging,' and 'premature death' are a function of the state-of-the-art of medicine at any given moment, then it is hard to imagine a solid basis for determining what health demands, expectations, or desires on the part of the elderly are reasonable or unreasonable (or what count as fair expectations of the young about their health as they age)."[11] Instead of basing our understanding of an acceptable lifespan on technology, Callahan proposes that we ground this understanding in nature. He then proceeds to define a "natural lifespan" and the related idea of a "tolerable death." Finally, he recommends forming a societal consensus about health care for different age groups based on these ideas.

According to Callahan, a natural lifespan is "one in which life's possibilities have on the whole been achieved and after which death may be understood as a sad but nonetheless relatively acceptable event." A tolerable death refers to "the individual event of death at that stage of life when (a) one's life possibilities have on the whole been accomplished; (b) one's moral obligations have been discharged; and (c) one's death will not seem to others an offense to sense or sensibility, or tempt others to despair and rage at the finitude of human existence."[12] The connection between nature and a tolerable death is that a tolerable death occurs only after a natural span of life has been lived.

Building on these ideas, Callahan presents an argument designed to show that persons are not entitled to receive publicly financed life-extending medical care once they pass the marker of a natural lifespan. The reasoning supporting this is that, if a person has lived a natural lifespan, then that person's death is tolerable. Moreover, if a person's death is tolerable, in Callahan's terms, then it is tolerable to allow death to happen by denying publicly supported life-extending medical care. Denial of care does not, then, depend on persuading particular patients that life-extending care is not in their interests or convincing them that they are not entitled to receive it.

Callahan's and Daniels' appeals to nature embrace several common themes. First, and most obviously, like Daniels, Callahan invokes nature to buttress an argument for imposing age limits on publicly financed life-extending care. For both, nature furnishes a general account of what expectations for health care are reasonable at different stages of life. Second, nature offers specific criteria for adjudicating competing demands for medical care. Certain demands are seen as legitimate, whereas others deviate from biological norms. Just as the idea of age-relative normal opportunity range sets limits to what we can expect from a just health care system, so our point in the natural lifespan determines, in part, what health care we are entitled to receive. Finally, the criteria for evaluating individuals' health care claims are global, rather than individual. Thus, for Daniels, knowledge of the species as a group determines what qualifies as normal or deviant for individuals, whereas for Callahan, a general understanding of the human lifespan grounds the determination of whether a particular person's life exceeds natural limits.

What is missing from both Daniels' and Callahan's accounts is a supporting argument for the view of nature each adopts. Thus, conceptions of nature play the role of basic concepts or first principles rather than rest on a solid foundation. Given the weight ideas of nature carry, the omission of a supporting argument is significant. What one would like to see is an enumeration of alternative philosophies of nature and some statement of the reasons favoring the approach each endorses.

How might Callahan respond to the above objections? In *Setting Limits,* Callahan writes: "I am not trying to read out of nature correct moral and social theories about aging and death. ...I want instead to use the term 'natural' in a different way, that of pointing to a persistent pattern of judgment in our culture and others of what it means to live out a life."[13] This passage ostensibly calls into question my earlier claim that conceptions of nature play the role of basic concepts. Instead, it now looks as if Callahan's appeal to nature is itself derived from a more fundamental social agreement.

In response to this, it can be said that, even if nature does not play the role of a basic concept in Callahan's theory, it nonetheless looms large in the idea of social agreement that does ground the theory at a deeper level. Labeling the outcome of social consensus "natural," and the span of life persons consider full a "natural lifespan," imparts to the idea of consensus the images and values we associate with nature. In this way, a particular social agreement appears sacrosanct; at the very least, it presents to us a far grander appearance than it otherwise would. Like the idea of nature, the idea of social consensus is then seen as authoritative and immutable, rather than created by finite and fallible beings like ourselves. The values inherent in such a consensus either go unnoticed or seem enshrined in nature herself. The unfortunate result is that Callahan's appeal to social consensus commands a dubious authority.

Thoreau's Perspective on Nature

As a way of developing a more systematic account of nature, I now turn to consider an alternative conception found in Thoreau. The alternative Thoreau puts forward suggests a reassociation of nature with individualism and a rethinking of the conceptions of nature sketched above. In the next section, I shall make more explicit the implications of this approach for the theories of age-group justice Daniels and Callahan present.

Thoreau first takes to the woods on Independence Day, July 4, 1845, and the theme of independence looms large throughout his text. It is, however, his own independence that Thoreau celebrates. In the early chapters of *Walden,* Thoreau exhibits a Cartesian propensity to doubt all beliefs and habits derived from custom or convention. Like Descartes, he uses the power of his own intellect to ferret out any views he holds that might spring from conventional sources: it "is a cleaver; it discerns and rifts its way into the secret of things. ...My instinct tells me that my head is an organ for burrowing, as some creatures use their snout and forepaws."[14] In the first chapter of *Walden,* Thoreau's concern to avoid unwitting reliance on others prompts him to give a detailed accounting of the construction of his house, including "all the pecuniary outgoes" he pays for additional building materials. "I give the details," he explains, "because very few are able to tell exactly what their houses cost, and fewer still, if any, the separate cost of the various materials which compose them."[15] He adds, "These statistics, however accidental and therefore uninstructive they may appear, as they have a certain completeness, have a certain value also. Nothing was given me of which I have not rendered some account."[16]

By presenting an account of what others have given him and so freeing himself of thoughtless reliance on others, Thoreau establishes the possibility of self-reliance. This possibility is realized at multiple levels:

1. By depending only on his own powers, he secures the physical "necessaries of life," that is, food, shelter, clothing, and fuel;
2. By relearning his own habits, he crafts daily plans; and
3. By sensing the passage of time in his own terms, he discovers that "morning is when I am awake."[17]

Ultimately, by distancing himself from others, Thoreau is able to shed conventional standards and create his life anew.

This last point emerges as a persistent and forceful theme throughout the book, and a major source for the account of nature Thoreau builds. "In our most trivial walks," he observes,

> we are constantly, though unconsciously, steering like pilots by certain well-known beacons and headlands, and if we go beyond our usual course we still carry in our minds the bearing of some neighboring cape; and not till we are completely lost...do we appreciate the vastness and strangeness of Nature. ...in other words, not till we have lost the world, do we begin to find ourselves.[18]

In Thoreau's terms, nature is paradoxically revealed as an individually crafted artifice. Thus, the opposite of nature is not artifice but convention.[19,20] Thoreau underscores the idea of nature as individual artistry by juxtaposing the bean field he plants with the large agricultural tracts planted by many, and by contrasting the simple home he builds for himself with the gaudy homes others purchase. Elsewhere, Thoreau compares nature to the herd: the solitary dweller in the woods is described in contrast to crowds heading for work on the "iron horse." To live naturally, then, means to "grow wild according to *thy* nature like these sedges and brakes, which will never become English hey,"[21] or to shift metaphors, what is unnatural is the conventional self we become among others: "I have long regretted that I was not as wise as the day I was born," laments Thoreau.[22] Reflecting on nature as artifice, Thoreau stands humbled. There are, he supposes, as many ways to fashion a natural life as there are people: "I desire that there be as many different persons in the world as possible; but I would have each one be very careful to find out and pursue his own way."[23]

These broad themes are richly suggestive of an alternative philosophy of nature. Notice how different Thoreau's approach is from the approaches considered earlier. First, although individual life arises out of biological life, what makes individual life natural is not conformity to general biological norms, but the eventual crafting of a life that goes beyond and distinguishes the individual as a unique, natural entity. Second, the cycles of growth, development, and decline, and the concomitant phases of youth, maturation, and old age that are common to all are underlaid by the individual's own sense of personal time. Individuals' experiences of time and maturation are seen as more real, more natural than any objective or public representation of them:

> My days were not days of the week, bearing the stamp of any heathen deity, nor were they mined into hours and fretted by the ticking of a clock; ...A man must find his occasions in himself.[24]

Finally, Thoreau suggests by his own example that individuals grow attuned to a natural life by dissociating themselves, at least temporarily, from society. Rather than turn to society or to a "heathen deity" to locate what life and death should be like, Thoreau recommends that individuals turn inward and shape authentic plans and values.

Implications of Thoreau's Views for Age-Group Justice

The philosophy of nature Thoreau suggests casts issues of age-group justice in a radically different light. Let me now attempt to draw out the implications of Thoreau's view for the ideas of normal species functioning, age-relative normal opportunity range, natural lifespan, and tolerable death. In the course of doing so, I will consider how shifts in the meanings of these terms affect the arguments about age-group justice in which they occur.

Consider, first, Norman Daniels' position and the idea of normal species functioning that plays a pivotal role in his argument. In Daniels' view, normal species functioning largely determines the range of opportunities that are normal for persons at each stage of life. A just health care system is responsible for ensuring that, so far as possible, deviations from normal opportunities are restored or compensated. If we accept Thoreau's individualist conception of nature, however, Daniels' account is seriously incomplete. What is left out is a description of normal functioning for a particular person, given that person's present life and future plans. In Thoreau's view, this would be an important foundation for Daniels' species-based account.

This change in the meaning of normal species functioning, in turn, changes what counts as a normal range of opportunities. A normal opportunity range will come to rest not only, and not even primarily, on normal *species* functioning, but instead, on some reconstruction of normal functioning for *particular* persons. The opportunities it is most important to restore or compensate will therefore differ as well. Accepting Thoreau's philosophy about nature implies that a just health care system cannot be based entirely on global or species criteria, but must also incorporate the histories and life plans of particular individuals at each stage of their lives. Thus, differences in the normal opportunity range for two 65-year-olds may mean that withholding life-extending care from one is just, whereas for the other, allowing death to occur would be premature or unnatural.

Whereas integrating individuals' perspectives enriches our ideas about normal functioning and age-related normal opportunities, leaving this perspective out is problematic in Thoreau's view. Doing so may obscure individuals' thinking about what they wish for their old age. At worst, it will shortchange persons by conveying to them that limited functioning and diminished opportunities are all they should expect in later life. This serves only to reinforce reduced functioning and hinder the prospect of finding personal fulfillment in later years.

It is important to note that individual lives are natural for Thoreau only to the extent that individuals are autonomous in shaping their own lives. Yet Thoreau cautions that "the mass of men lead lives of quiet desperation."[25] This suggests that a just health care system faces the considerable task of eliciting individuals' underlying values and goals, rather than simply basing patient care on individuals' expressed wishes.

Let us next turn to consider Callahan's definition of natural lifespan. Recall that Callahan supports forming a social consensus about what counts as a natural lifespan, and then designing a conception of just health care consonant with this. In contrast to Callahan, Thoreau would undoubtedly emphasize individuals' unique and subjective experiences of their place in the lifespan. For example, at age 35, one individual may be settling into midlife, with a secure job, several offspring, and a newly purchased home, whereas another person of the same age may be just beginning to face finding a mate, embarking on a career, and gaining financial security. Thoreau's emphasis on individuals suggests rather strongly that the socially defined natural lifespan Callahan proposes is inadequate. What is needed is an account that rings true for individuals and takes heed of individual differences. A social consensus alone can never determine where individuals are in their personal histories or whether their histories are near a point of natural closure. Global criteria can at best dimly illuminate these questions. Only the authors of life plans are able to say with surety how far along they are in carrying out particular plans. Moreover, different persons are likely to locate meaning and value at different places in the lifespan. Thus, the meaning and value of later ages will vary and depend on what individuals deem important in their lives.

Just as, for Thoreau, individuals' sense of personal time and narrative inform the meaning of natural lifespan, so too what qualifies as a tolerable time to die depends on individuals' experience:

> Let us settle ourselves, and work and wedge our feet downward through the mud and slush of opinion, and prejudice, and tradition, and delusion, and appearance, that alluvion which covers the globe...till we come to a hard bottom and rocks in place, which we can call *reality,* and say, this is, and no mistake...Be it life or death, we crave only reality. If we are

> really dying let us hear the rattle in our throats and feel cold in the extremities; if we are alive, let us go about our business.[26]

Thus, although Thoreau might agree with Callahan that a tolerable death occurs only after a natural lifespan has been achieved, for Thoreau a *natural* lifespan reflects the individual's own sense of fitting closure.

As before, Thoreau's stance makes leaving out individual perspectives problematic. Doing so poses obstacles to individuals' efforts to figure out what they would like their life and death to be like. At worst, leaving out individuals' viewpoints and defining natural lifespan and tolerable death by social consensus create the impression that after a certain age all important tasks have been discharged and life is superfluous. The implication of this way of thinking is that life in old age can no longer occur with dignity or pleasure. By contrast, viewing a tolerable death in individualist terms implies that such a death has very little to do with whether the timing of death conforms to social expectations. A tolerable or natural death also has little to do with whether "artificial" means are used to keep persons alive. Instead, a good death for an individual depends on the extent to which the individual can craft his or her own death and others can be responsive to his or her wishes.[27]

Conclusion

The forgoing remarks make evident that appeals to nature play a crucial role in current debates about age-group justice. They also intend to show how our most fundamental ideas about nature admit of alternative interpretations. Failing to attend to this is by no means a new phenomenon. Since its inception, the field of geriatrics has been prone to relying on ideas of nature and biology, and treating these ideas as sacrosanct. When scientists and physicians first devoted themselves to the study of senescence in the nineteenth century, they intended to cast off religious dogma and metaphysical explanation, and turned to nature and biology to discover The Truth about aging. Thus, the founding fathers of geriatrics wrote about old age as if they were formulating the pristine laws of its nature.[28]

Today, ideas about nature still infuse our thinking about aging. It should come as no surprise, then, that conceptions of nature underpin ethical argument about health care for the elderly. Rather than attempting in vain to purge our thoughts of these ideas, we should strive to incorporate them in a more open and reflective way.

References

[1]H. D. Thoreau (1960) Walden, in *Walden and Civil Disobedience* (S. Paul, ed.), Houghton Mifflin, Boston, MA, p. 95.

[2]D. M. High (1978) Is natural death an illusion? *Hastings Center Report* **8,** 37–42.

[3]A. MacIntyre (1966) *A Short History of Ethics*. Macmillan, New York, NY.

[4]M. G. White (1981) *What Is and What Ought to Be Done: An Essay on Ethics and Epistemology*. Oxford University Press, New York, NY.

[5]N. Daniels (1981) Health-care needs and distributive justice. *Philos. Public Affairs* **10,** 146–179.

[6]N. S. Jecker (1989) Towards a theory of age group justice. *J. Med. Philos.* **14,** 166, 655–676.

[7]N. S. Daniels (1988) *Am I My Parents Keeper? An Essay on Justice Between the Young and the Old.* Oxford University Press, New York, NY, p. 74.

[8]N. Daniels (1985) Just Health Care. Cambridge University Press, Cambridge, UK, p. 30.

[9]D. Callahan (1987) *Setting Limits: Medical Goals in an Aging Society.* Simon and Schuster, New York, NY.

[10]Callahan, pp, 54,56.

[11]Callahan, p. 56.

[12]Callahan, p. 66.

[13]Callahan, pp. 64,65.

[14]Thoreau, p. 68.

[15]Thoreau, p. 33.

[16]Thoreau, pp. 41,42.

[17]Thoreau, pp. 61,62.

[18]Thoreau, p. 118.

[19]S. Gadow (1987) Death and age: A natural connection? *Generations,* **Spring,** 15–18.

[20]S. Gadow (1989) Natural death, in *Rounds: Selections from the First Five Years* (K. M. Stephens, ed.), University of Texas Medical Branch, Galveston, TX.

[21]Thoreau, p. 68.
[22]Thoreau, p. 68.
[23]Thoreau, p. 49.
[24]Thoreau, p. 78.
[25]Thoreau, p. 4.
[26]Thoreau, pp. 67,68.
[27]M. Battin (1983) The least worst death. *Hastings Center Report* **13,** 13–16.
[28]T. Cole (1992) *The Journey of Life: A Cultural History of Aging in America.* Cambridge University Press, New York, NY.

Paying the Real Costs of Lifesaving

Paul T. Menzel

What is it worth to save people for this or that many years of life? Sometimes we pose that question by asking more crudely what a life is worth, perhaps even what a particular elderly person's life is worth. In any case, of course, we need to know what it really costs to save a life. Then, with some idea of both costs and worth in hand, we proceed to decide whether we should spend what it costs.

In this chapter, I will first clarify what the economic question, "what is a life worth?" means, arguing that in health care policy it is not precisely the right question to ask. Then I will explore some commonly ignored dimensions of the cost of saving life, concentrating on two that are especially important in the case of lifesaving decisions for elderly people—the later health care expenditures and added pension payouts that prolonging life often involves. Finally, I will respond to the main issue to which all of this points: what costs we should be willing to incur to save a life.

Throughout, I will have in mind primarily a context of saving the lives of elderly people. I will use one kind of decision frequently as an example: saving the lives of elderly residents of nursing homes. Make that even more specific: saving their lives by using nominally very inexpensive medical procedures, such as vaccine for influenza or penicillin for pneumonia. The pneumonia case is proverbial and common. The flu vaccine case is less well known, but equally significant. One noted economic analysis of influenza immunization

Aging and Ethics Ed.: N. Jecker

for the US Medicare population portrayed direct medical costs as only $13 per year of healthy life saved.[1] At that price, we wonder how any sane Medicare administration could fail to cover routine immunization. According to the same study, including the costs of later unrelated health care (but not long-term nursing home expenses) for the patients whose lives are prolonged increased the cost to $800 per year of healthy life, still a very good bargain as medical lifesaving goes. The inclusion of longer years of Social Security payments and nursing home bills, however, may bring costs to a level where the argument for passing over vaccination becomes plausible. Perhaps it is, in fact, the worry about additional years of medical expenses, nursing home bills, and Social Security benefits that actually explains why the Medicare administration and Congress have been so reticent to pay for routine vaccination.

The Wrong Question: What Is a Life Worth?

Some sort of monetary valuation of life seems to be an integral part of modern life. With the availability of a great number of possibly effective, but often (statistically, at least) costly means of preserving life and preventing death, we repeatedly trade lifesaving off against other things, and it is money, of course, that mediates the trade-offs. In the case of the elderly, if we ever refrain from employing some lifesaving care, acute or preventive, we presumably think that the measures we forgo "cost too much"—referring, of course, to the perceptually more valuable things we could use the resources for elsewhere. Also, it is the actual lives of older individuals that we sometimes thereby really lose. For purposes of discussion, suppose that the cost we think excessive computes out to something over $500,000 per life saved—two $250,000-plus liver transplants, say, each with an estimated 50% chance of long-term success, or 1000 at least $500 CAT scans that ultimately help us save a total of one life.

Economists often have talked about such cases in terms of an actual dollar figure for the value of a life—$500,000, say, under the suppositions just described. The advantage of that way of thinking is that we thereby tie together the whole conceptual package of resource trading. Using a dollar figure as "the value of the life" of an older person, we can say that, if we spend only that and no more to save the life, that is because incurring any greater cost would exceed the value produced. "Value for money," after all, is what we are after in any of this resource trading. It is not that in some *independent* sense the value of the life we might save is less than what we spend to save it, but only that what we have decided to spend to save it *marks it* as having the value we then say it has.

There is even considerable moral sense in such a way of thinking: we do not need to spend more than that amount to save a life if the older people whose lives end up not being saved, themselves having representative preferences, presumably would have gambled at an earlier time. If they, too, would not have been willing to devote resources to the lifesaving enterprise at more than the dollars-per-life-saved rate we call "the money value of their lives," then their presumed prior consent to take these risks morally justifies economic limits to the health care they ought to be provided.[2]

It is hardly strange, then, that economists, at least, have frequently endorsed such talk about the actual money value of people's lives. They have typically called the particular approach just described the "willingness-to-pay" method for establishing that value.[3] However, though in theory this approach may seem eminently plausible, it has hardly gained moral acceptance in addressing people's health care, including care for older patients. The basic problem is simply that in the end the world is such a different place for a loser than it is for a winner. Suppose one refuses to pay more than the $500 risk-reduction price for one of those 1000 CAT scans, and then one later (maybe only very shortly later) dies because of that decision. Of course one has in some sense consented to what happened, but did one ever say anything remotely like "$500,000—no more—is the value of my *very life*," the life that *after* the fact is irretrievably lost? How does anyone using the willingness-to-pay approach get thrown from an initial trade-off between money and risk, a trade-off we actually do (or would) voluntarily make, to the value of a real, irreplaceable life? "Her life [now lost] was worth only $500,000"—after she is dead, should we actually think things like that?

One critic among the economists themselves has elaborated these doubts at length. John Broome claims that in principle only valuations of life made directly in the face of death are correct.[4] They must be used to evaluate what actually happens because of a policy; any prior acceptance of risk is only a valuation of the subjective expectations the policy creates. Such a valuation is useful only to the degree that it approximates its subsequent counterpart, but most prior valuations of life and death in low risk contexts yield significantly *lower* monetary figures than valuations made closer to anticipated actual death. Mishan, for example, quips that he has yet to meet a colleague "who would honestly agree to accept any sum of money to enter a gamble in which, if at the first toss of a coin it came down heads, he would be summarily executed," whereas it would be a very strange lifestyle indeed in which we were not willing to take low risks to life to save resources.[5] In the justification of future-looking policy, then, Broome claims, the low risk valuations are "worthless, and...known to be worthless at the time" they are made.[6]

Broome's objection contains both a very correct point and a badly mistaken one. The mistaken claim that Broome and some others try to pull from this is that, since the value of the life that *is* lost has nothing like the finite limit of $500,000, no policy that limits the use of lifesaving resources can possibly be justified. After all, it is said, if the value of the life that might have been saved cannot be said to be worth less than what it would cost to save it, how can refusal to pay those costs be justified? If it is worth more, that greater amount, whatever it is, should be spent.

For *health care policy,* though, this is flatly mistaken reasoning. In that context, we cannot assume, as Broome does, that the later perspective of an elderly individual, for example, immediately in the face of death is the correct one from which to make policy decisions. That later, in-the-face-of-death perspective may be correct for the legal process of estimating the value of life for wrongful death awards, for example; precisely what we are trying to do there is compensate people for losses actually incurred. However, why assume that for all decision contexts we must judge from that later, loss-at-the-time vantage point? Valuations of life immediately at the time of peril are not, in fact, the proper standard against which *health care* resource decisions have to be measured. The reasons for using a younger or older person's more prior perspective relate directly to the nature of the health care economy and the human need for people to get control over the resources of their lives. In modern medical economies, most people either subscribe to insurance plans or are covered by public health care spending. Once patients are insured, whether in private or public arrangements, they as well as providers naturally get very strong incentives to overuse care and underestimate opportunity costs (the value of alternative uses of the money). Why should we not address the problem of controlling the use of care in the face of these incentives at that point in the decision process—insuring—where the cost-expansion trouble fundamentally starts? If we, voters in a democracy or members of an insurance pool, decide against the incremental tax or premium necessary to provide coverage for liver transplants, for example, then controlling costs by rationing out those procedures issues fundamentally from our own will. That will be true even if the coverage passed over is more specifically age targeted—transplants for people over 65, for example. If at a later time or older age we find ourselves in liver failure but are denied a liver transplant just like other voters or premium payers in the same pool, it is our own policy that is being used against us; we are simply living by our decisions to control the scarce resources of our lives.

To be sure, sometimes there are reasons for not allowing people to precommit themselves to policies they might later wish to avoid or reject. However, the best reason—that a later decision is likely to be better—simply does not hold in health care rationing contexts. In the later context of an insured, ill, and older patient, people's capacities to control the resources of their larger lives are very sharply diminished. If as either communities, individuals, or both, patients want to retrieve that control (and not just hand it over to someone else), they have no alternative but to make prior health policy decisions. Any view that rejects people's power to exercise control over what they know will be their desires in later life as patients in peril would hardly respect their autonomy as persons in any full or mature sense. What conclusion can be drawn from this? Broome's rejection of prior willingness to take risk as a basis for holding people to economic limits on what is spent to save their lives is mistaken.

Nonetheless, a correct point is harbored in his and everyone else's suspicion about how the real worth of an actual life could be anything like a finite monetary value. It is only, however, a verbal point: after-death or in-the-face-of-death judgments must be our point of reference if what we are expressing is properly to be called *the value of an actual life*. However, those who argue that there should be limits on what is spent to save life can simply admit this. Prior consent to risk—agreeing to a policy that rejects some life-saving measures to have resources left for other goods in life—is really only a way of pricing *safety,* not life itself.[7] The term "the value of life" was anyway "almost a joke, a bit of gallows humor" inevitably attractive to economists who routinely work with tragedy-loaded scarcity.[8]

By admitting that "what a life is worth" is not what is being revealed by prior consent to risk, defenders of limiting what we spend to save lives have lost nothing of substance in their argument. Suppose an elderly patient previously consented to a lean health plan that now leaves him or her on his or her deathbed, say, and his or her risk-for-money trade-offs point to a limit of $500,000 per life saved. If we call that an actual "value of his or her life," are we any *better* situated to explain why all the stops should not be pulled out now to save the patient? We do not have to use any "$500,000 value of life" language at all. The essential point is that the patient consented to something beforehand. Saying or thinking, "See, $500,000 is all your life is worth; you said so," adds nothing morally significant. What we need—all we need—is sufficient reason for people to bind themselves beforehand for contingencies later. In health care, given the incentives created by insurance, we have that.

"What is an elderly person's life worth—an actual *life* that might well be lost?" is the wrong question. "What will that person ahead of time agree to spend on this (or that) way to save lives?" is the important one.

What Does It Really Cost to Save a Life?

What are we really spending when we save the life of an older person? We have to get clear on that if this whole business of trying seriously to decide what to spend is remotely to make any sense. Suppose, for example, that we fund the immunizing of nursing home residents with flu vaccine. We tend to think of the cost of that lifesaving program as just the money we have to lay out directly to fund it, but the matter is much more complex.

Take a somewhat different case first as a comparative, illustrative example. We fund a lifesaving antismoking program and think of the nominal cost we have to pay for it directly. Perhaps the program also saves some money: it allows people to live further money-earning years, and it obviates the need for the tobacco itself, for repairing the fires that smokers cause, and for the medical care that would have been spent caring for their smoking-related diseases. It can begin to look like a real bargain, even a money-*saving* way of saving lives.

However, then along comes a shrewder cost analyst. In the long haul, she says, reducing the incidence of smoking in the population does not save the government or other people anything at all, for there are two huge costs of lengthening the lives we have saved: later unrelated health care expenditures and longer pension payouts. The typical nonsmoker hardly lives perfectly healthily to 85 only to get killed instantly; he or she incurs the typical and often costly diseases of older age, and he or she draws additional years of Social

Security benefits. In the long run, smoking, it turns out, does not really cost much money at all, and stopping it, although saving lives, does not save money.[9]

Such a factual analysis may surprise many people, but we should note that it hardly poses a serious objection to noncoercive, reasonably designed antismoking programs; we have realized that they may not save us money, but whatever small net expense we incur seems well justified by the value of the life itself that gets prolonged. The matter can be disturbingly different, however, with other health measures, particularly those for elderly people that occur both at the end of their earning years and when they are about to begin running up the bigger expenses of the more common diseases of old age. Go back to our nursing home resident influenza immunization program, or other simple and apparently inexpensive health care measures for elderly populations. If we use penicillin to stop the pneumonia, for example, the total costs ultimately incurred include not only the $50 for the penicillin shot and its administration, but all the additional medical and long-term care expenses and pension payouts in whatever years of life we save that person for. Unless resource trade-offs are totally irrelevant and we ought never to limit effective health-producing care of any sort, should we not know what we really will be paying before we decide to save lives?

In the next section, I simply want to explore more critically whether these two often unacknowledged costs of lifesaving—later pension payouts and unrelated medical expenses—really in fact *are* actual costs. In a subsequent section, I will ask whether, if they are real costs, we *should count* them in conceiving what it is we would be paying to save a life.

Future Health Care Expenditures

For future health care expenditures incurred by extending the lives of the elderly, "is it a cost?" gets answered quickly: later health care expenditures, even for medical problems unrelated to the problem we might now be alleviating, are real costs of our present lifesaving measures. I can imagine only two objections to such a claim, neither of them on target.

(1) One economist, Louise Russell, refuses to count such expenses in her economic analyses of various preventive health measures. Added years' medical expenses are irrelevant, she says, because they "are one of the indirect consequences of the health gains from a program. ...They are not an addition to health effects."[10]

Russell fails to explain why the distinction between "health effect" and "indirect consequence" should matter. Are indirect costs that come as a consequence of a decision any less real? Perhaps her decision to ignore the indirect consequences in an economic analysis is justified by a hidden assumption about which of the admittedly real costs we *should* count: the moral distinction between intended and merely foreknown effects. This well known doctrine of "double effect" says that, though we *foresee* the later unrelated health care expenses of your added life when we save you from pneumonia with penicillin, we surely do not *intend* to create those expenses. We do not even intend them as a means to what we aim at (your longer life). They are only "side effects."

If this is Russell's argument, it does not help us much in our query. We would still have to admit that these later health care expenditures *are* real costs. Also, even if we admit this but only want to claim that morally they should not be counted, we would still be on slippery ground: the moral significance of the difference between intended and merely foreknown effects is itself notoriously debatable.[11]

(2) Alternatively, one might argue that, though later health care expenditures are real costs, now, for example, in the current decision about whether to administer penicillin to a 75-year-old, is not the time to count them. They should—and will—be counted when later decisions are made about whether to use the measures that incur them. People will be deciding then whether the human benefit of those measures justifies their costs.

This may seem to make sense, but note how excluding these later statistically predictable medical expenses from assessments of lifesaving programs would require unusual and incredibly cumbersome cost-counting procedures. To be consistent, we also would have to exclude from the *benefit* side of our current assessment all the life and health benefits of the later care, counting only the benefits achieved *between now and then.* In the world of statistical estimates of what life a particular program buys, how could we separate out those two temporal categories of benefit? It would seem much more sensible just to assume that any later health care will buy sufficient benefits to be worth its cost then, and consequently count now both the costs and benefits of that later care.

So both reasons for not seeing later health care expenditures as real costs fail. They are real costs that we are paying with a current decision to save a life.

Future Pension Payouts

With respect to future pension payouts, the problem is much more complex. A virtually standard assumption among economists is that pension benefits to the elderly are transfer payments and, therefore not overall costs from any fully society-wide point of view. Assume that the tax-transfer scheme does not create any serious disincentives to work and invest, and does not incur significant administrative costs. Then, though someone through taxes or premiums pays such benefits to someone else, "society"—everyone, all of us together—incurs no net cost at all. When older people die, they have lost a pension benefit, but the rest of us have saved a roughly equal expense; when the lives of older people are saved, we have incurred the expense of later pension payouts, but they receive that as an equivalent benefit. Either way this contrasts with consumption, which involves real uses of goods and services.[12]

This view is correct if we are comparing two courses of events among the numerically same population. Either one transfers to another and the other gains as much in value as one might have kept, or one keeps what one has and the other gets along without, though there, too, the other is still around. I do not want to challenge this traditional economic explanation of why pensions are virtually costless for the typical set of circumstances that it assumes: people are staying alive whether or not the transfer is made. I do want to challenge this view for other settings, however.

Things look very different when we compare courses of events in which the *number* of people changes. That is exactly what happens in lifesaving programs. We are comparing (a) a course of events in which one is alive to receive a payment from others, with (b) a course in which the others save that expense, but not because one is "doing without"—one is just not around at all.

Compared to the second, is the first a "cost"? To the others, of course, it is (so it is clearly what economists call an "external cost"), but that is not our concern here. Is it a cost from the wider, total, social point of view? There is an obvious sense in which it is neutralized by being a benefit to one, yet only "a sense." In the second course of events in which one dies, there are fewer people among whom the goods and services will be spread. Someone somewhere down the economic line is going to have more there than in the first situation, without anyone in the second getting less. In this respect, *b* is a gain compared to *a,* and *a* is a cost compared to *b.*

The easiest way to describe this respect is probably to say that per capita net income is lower in *a.* This does not refer to one's own per capita income *per se,* or to another's, but to per capita income generally. The frame of reference is still completely society-wide. So in one genuinely societal sense, pensions are real costs. Whether it is per capita or aggregate income that ought to be considered is then itself the important and open question. Many economists perhaps focus more on aggregate than per capita benefit because of the impersonal sense of value with which they typically work. Most of us, however, think more in terms of a per capita perspective. Particularly in a resource trade-off situation, people are trying to decide whether their lives will be better or worse off by incurring or not incurring this or that lifesaving expense. It is the per-person likely benefit, not some more abstract total good, that people are trying to keep in focus. They are, of course, taking due account indeed of whether they might be the person who is or is not around as the policy in question plays itself out, but that hardly requires an exclusive focus on some aggregate of benefit that disregards the number of persons among whom it is divided.

For purposes of making resource allocation decisions, therefore, later pension payouts seem to be a real cost incurred in saving the lives of older persons. There is a remaining way in which some people might try to save the costless transfer payment view, however, even if we are focusing on per capita costs, but in the end I think that it, too, is very problematic. Suppose the missing person in the second course of events above would have injected additional goods and services into the first course had he or she lived: some distinctive "labor" in old age—cheering up children, for example, or teaching others the subtle lessons of gracious appreciation. He or she pulls out more pension benefits in his or her added years of life, but creates fully compensating, equivalent goods; they are just hidden.

However, that is a tall order—*equivalent* goods. By hypothesis, these goods are not created through paid economic productivity, nor are they hidden in the "value of life" for added years that will already be counted on the intangible benefit side when we subsequently decide whether or not the cost is worth paying. These goods have to reside in some other kind of value in the pensioner's longer life. Without them, the pension payouts of added years of life still emerge as real costs.

I conclude that future pension payouts as well as later unrelated health care expenditures are real costs of lifesaving. In the nursing home/flu vaccine case, for example, we are incurring much greater costs than the nominal money for the vaccine and its administration. Per year of life prolonged, modest estimates might easily be another $2000 of unrelated medical expenditures, $7000 of Social Security payments, and $20,000 in nursing home fees. We could easily approach costs of $30,000 for every year of life we save with flu vaccine. This puts it in the same ballpark of admittedly cost-controversial care as kidney dialysis and heart and liver transplants.

Ought We to Count All the Real Costs?

There are, however, a number of arguments and certainly some widely held intuitions that tell us we should not count some or all of these two kinds of costs as we go about considering what to spend to save life. I have noted already that one argument against counting later health care expenditures that are admittedly real costs fails. In this section, I shall pursue two further arguments.

Necessities and Rights: The Food-and-Clothing Parallel

If someone proposed counting future food and clothing costs in deciding whether to save a life, our reaction would be, "You cannot count those; they are just part of what anybody needs to live!"? This response does not deny that food and clothing are actual costs. It only makes a moral claim about them: when people continue to live, they should not in any way be held accountable for using up financial resources for the bare essentials of life. Anyone who lives has a right to those.

Put the matter this way for health care expenditures. Suppose an elderly person lives longer. Thus, alive, that person certainly cannot be expected to reduce or eliminate the consumption of bare essentials. However, later medical care is just as much a necessity as food or clothing, so if that individual should not be "charged" for the food and clothing he or she uses in living longer, so also he or she should never have to justify the later use of medical resources.

This argument, I suspect, is at the core of our common reactions to this cost-counting issue, but it is beset with problems. Admittedly, one has a right to minimal food, clothing, shelter, and health care in the time during which one is *already assumed to be living,* but why must we therefore ignore these items' costs when the matter at issue is explicitly the *extension* of one's life? Admittedly, food is a necessity to which people have rights partly or even largely because it is essential to their living longer, not merely to enjoying the life they would have in any case. Still, is there not a difference between rights to things when being alive is assumed as a background condition and rights to those things when life extension is the issue? If we had a distinct shortage of food, for example, would we not see to it first that people who were going to live a considerable time anyhow got sufficient food to avoid lingering and debilitating malnutrition? Would we not think it crucial to count food consumed in figuring the cost of saving people's lives?

So the parallel-with-food-and-clothing argument, although cogent, is sharply limited.

1. We might not see all of the later health care whose expense is at issue to be a clear necessity. The argument demands that later medical care be regarded as "essential" or strictly "minimally decent"—that is, as care to which people have virtually as strong a right as they have to food. In the current day, when many people are giving various categories of care low enough priority to regard them no longer as strict necessities, it is not surprising that we are attracted to counting the health care expenses of added years of life.
2. The more that basic necessities themselves are seen to be scarce, the weaker the food-and-clothing argument gets. If there are sufficient resources to provide everyone with minimal food and clothing, even those who might live longer, we should not consider the cost of such essentials. On the other hand, if we already regard such essentials as themselves scarce, we are hardly attracted to the argument at all.
3. Finally, we are starting to see the life extension from deliberate decisions for lifesaving programs as significantly different from the natural continuance of people's lives. In fact, the whole business of finding ourselves in a resource scarcity situation, where we are really beginning to make priority decisions between lifesaving measures and other good things in life, just constitutes precisely a situation where we single out decisions to extend life from decisions in which either way the affected people are going to be alive. Any serious

economic inquiry in a larger rationing context already has us seeing a difference between people just continuing to live and being deliberately saved. The first two limitations above may not apply clearly to our situation in the United States of the 1990s, but this one catches us in the real cultural and economic shift in which we now find ourselves.

How does the food-and-clothing objection then finally stack up as an argument against counting later health care expenditures? It falters as our perception of health care as a common essential in added years of life diminishes, as we view health care resources themselves as scarce, and as we see deliberate life extension to be different from life's natural, in-any-case continuance. Especially because of the last, the parallel-to-food-and-clothing argument for excluding later health care expenditures now virtually falls on its face. Although the objection is bothersome, it is hardly logically persuasive.

In the case of pensions, we face a similar, and similarly weak, parallel-to-food-and-clothing argument. Here, the argument gets made by saying that pensions are older people's basic right. If society sets up a nonannuity pension scheme because it thinks older people have a right to be assured such a base of support regardless of their private arrangements, why should it "charge" pension costs against them when they live longer and draw out what they have a right to draw out?

However, here we quickly come back to the same problems in the argument. One may have a right to a pension benefit for the time when one *is* alive, but that hardly commits us to saying that one has a right to have one's life *extended* without regard to the impact one has on others by drawing the benefit. We may not see much of a difference between these two sorts (degrees) of rights in the case of food, clothing, and really basic medical care. Insofar as pensions have been set up in order to provide for such equally basic needs as these, they may carry with them some of the force of the parallel with food and clothing. The parallel will deteriorate, however, to the degree that a pension program's benefits are not really "needed" by its recipients. Currently in the US, in fact, that consideration is very much at issue; Social Security benefits are often no longer regarded as going primarily to those in need.

Thus, whereas both later health care expenditures and pension payouts might be in some sense older people's right, it is not clear that they should be excluded as we count up costs to decide whether lifesaving is worth what it costs. In the absence of any clear, reasonably persuasive argument against counting what we have already admitted is a real cost, we should count all these items.

Benefits Already Paid for

Finally, consider pensions as benefits already paid for. The paradigm here is an annuity pension, into which one pays roughly as much as one is statistically likely to draw out later in life. The fund probably will require one to pay higher premiums the longer one is expected to live. If one paid in what the fund asked one to, and if the fund did not charge one more when it might have, then one would seem to have as much right to additional years of payouts as one did to earlier years. If one lives longer as a vaccinated nursing home resident, only in a weak sense are one's higher pension benefits external costs passed on to others—one has defrayed them already in the pension purchased. In an exceedingly robust sense, one is *entitled* to them. Why, therefore, should an influenza vaccination program have to carry these costs into the debate over whether it should become policy?

This argument is familiar, but its limitation is also clear: it works straightforwardly to exclude additional pension payouts from our cost counting only when they are part of annuity pensions. Neither British old age pensions nor American Social Security pensions are. They admittedly have some "collective annuity" dimension to them: since people paid into the system all those years, they have a right to get out something close to what was projected for them, but that projection is not based straightforwardly on what any particular person paid in, so people can hardly expect others to ignore the higher payouts one might run up. If one just is going to be alive, then clearly one has a right to whatever schedule of benefits is operative at the time, but this does not tell us that such benefits should not be counted as a cost in the different situation where whether one will live longer is precisely the matter at stake.

So this argument, too, fails. We have not been able to articulate a persuasive case for excluding health care expenditures and pension payouts to the elderly as we consider what costs we might be willing to pay.

Should We Pay the Real Costs?

We now have a clearer idea of what the real costs of lifesaving for older persons are, and we have decided that virtually all major categories of these costs should be counted as we enter the consideration of what to spend to save lives. However, now we face the question we have been leading up to all along: Should we save the lives of elderly nursing home residents with flu vaccine at the cost of nearly $30,000 per year of life saved?

"Age Rationing" Is Not the Issue

Putting the question this way, of course, may seem to be loading it against lifesaving care because it is the *elderly* whom we are asking whether to save. In our case for discussion, lifesaving's cost is as high as it is because of factors peculiarly associated with old age—Social Security payouts for all subsequent years, nursing home care that is rarely needed by younger patients, and higher than average anticipated annual medical expenses given the fact that the person is in the highest medical cost period of life. The benefits, too, seem less; at this relatively late point in life, we will not be saving people for that many more years. All this may seem to load the dice against lifesaving because people are old, and is that not age-discrimination?

This is a complicated issue. It is clear, first of all, that any decision not to pay this cost to save such lives is not a pure *age* rationing. Pure age rationing would not confine its knife to lower benefit, higher cost-per-benefit care that happened to present itself in old age; instead, when one reached the appropriate age, that fact itself would determine that life-extending care would drop in priority. It is not surprising that we resist proposals for such rationing; they seem to imply that, even when segments of life bought after a certain age are rather long and personally well appreciated, they are still less valuable. However, that seems incorrect, even from earlier (or time-neutral, lifelong) perspectives; in trying to make tough saving and allocation decisions for the resources of our whole lives, it is doubtful that we will just throw all life-extending care after a given age into the same low priority bag. "If I get there," we think, "maybe it will be my most appreciated time in life." If we are going to ration care at all in old age, almost all of us would prefer to ration by the relevant possible characteristics of care that may attend old age, not by age itself.

However, does not a decision against spending to save life because of the various typical characteristics of old age that make the cost–benefit ratio so high constitute a more subtle kind of discrimination against the elderly? Indeed, I could see that it might were it not for the fact that the selection of this high-cost-per-benefit care as some of the first care to limit is likely to emerge from an entirely age-neutral perspective. To control the resources of our lives, almost all of us will certainly be willing to bind ourselves to limits on the lowest benefit-per-cost items in the system. To be sure, such potential for precommitment is abused by anyone who does not seriously reflect on his or her future. Yet how can moving exclusively to desires held in old age, ignoring the distortion of cost–benefit judgment that happens in the immediate situation once we are insured, constitute any kind of satisfactory antidote? Indeed, we should discount the choice of a 30-year-old not to invest in any policy that provides care that prolongs life in his or her eighties; living an extra year at 80 may seem unimportant to one at 30, and that is reason enough to take a 30-year-old's perspective on old age policy with a grain of salt. On the other hand, none of these observations are reason to take the octogenarian's word as controlling, either.[13]

It should be kept in mind, too, that people who are wisely trying to manage available resources for their lifetimes will clearly not commit themselves to doing without palliative or chronic care no matter how old they may become. We might discount the relative priority of lifesaving rescue care as old age sets in, but we hardly will condemn ourselves to misery and lack of care regardless of how old we might be if we see ourselves as then being alive anyhow. Unless palliative care is incredibly expensive or the need for it very short-term, virtually no foresightful subscriber or voter will cut it short in any prior allocation decision. Long-term and palliative care—*all* the nonlifesaving care of elderly patients, in fact—stay as high priority as they are at any age. People who worry that rationing care for the elderly or terminally ill will lead us to deny them palliative or chronic care simply forget the prior consent rationale for rationing.

Conclusion

There is no general or abstract way to decide whether $30,000 per year of life saved at 75 is too high a cost to pay for penicillin for pneumonia or vaccine for influenza. Everything will depend on all the other things we want to do with the resources of our lives. What is clear is that people of integrity, appreciating all the ages they might live into, will not hide their heads in the sand about what the real costs of lifesaving are. There is seldom a free lunch; lifesaving care in old age certainly is not one of them.

We also can no longer bury our heads in the sand and pretend that we do not have to, or should not, make hard trade-offs in our lives among relative priorities. In effect, we have already decided to engage in the difficult trade-off game in adopting a public program like Medicare for the bulk of care for the aged. Since then, the context of our thinking about medical care for the elderly has necessarily had to become what we might call "congressional"—seeing ourselves as legislators, responsible for the use of resources over our larger lives. Whether to pay the real cost of saving life in old age is always an open question, but facing up to the real costs of what we are doing is not. We must face up.[14]

Notes and References

[1]M. Riddiough, J. Sisk, and J. Bell (1983) Influenza vaccination—cost-effectiveness and public policy. *JAMA* **249,** 3189–3195. A year of "healthy life" refers to a "quality-adjusted life year," or "QALY"; *see* P. Menzel (1990) *Strong Medicine: The Ethical Rationing of Health Care.* Oxford University Press, New York, NY, pp. 79–96.

[2]This role of consent is spelled out much more carefully in Menzel, *Strong Medicine,* pp. 10–15 and 22–36.

[3]The two seminal pieces that launched this model were T. Schelling (1968) The life you save may be your own, in *Problems in Public Expenditure Analysis* (S. Chase, ed.), Brookings, Washington, DC, pp. 127–62; and E. Mishan (1971) Evaluation of life and limb: A theoretical approach, *Journal of Political Economy* **79,** 687–706. Two important later treatments are E. Mishan (1985) Consistency in the valuation of life: A wild goose chase?, in *Ethics and Economics* (E. Paul, F. Miller, and J. Paul, eds.), Blackwell, Oxford, UK, pp. 152–167; and H. Leonard and R. Zeckhauser (1986) Cost–benefit analysis applied to risks: Its philosophy and legitimacy, in *Values at Risk* (D. MacLean, ed.), Rowman and Allenheld, Totowa, NJ, pp. 31–48.

[4]J. Broome (1978) Trying to value a life. *Journal of Public Economics* **9,** 91–100. J. Broome (1982) Uncertainty in welfare economics, and the value of life, in *The Value of Life and Safety* (M. Jones-Lee, ed.), North-Holland, Leiden, Netherlands, pp. 201–217.

[5]Mishan (1985), in *Ethics and Economics,* pp. 159–160, *see* Note 3.

[6]Broome (1978), in *Journal of Public Economics* **9,** 95, *see* Note 4.

[7]Mishan (1985), in *Ethics and Economics,* p. 165, *see* Note 3.

[8]D. Usher (1985) The value of life for decision making in the public sector, in *Ethics and Economics,* p. 168, *see* Note 3.

[9]K. Warner (1987) Health and economic implications of a tobacco-free society. *JAMA* **258,** 2080–2086.

[10]L. Russell (1986) *Is Prevention Better than Cure?,* Brookings, Washington, DC, pp. 35,36.

[11]J. Bennett (1985) Morality and consequences, in *The Ethics of War and Nuclear Deterrence* (J. Sterba, ed.), Wadsworth, Belmont, CA, pp. 23–29. For

a defense of the now generally minority view in philosophy that the distinction is directly morally relevant, *see* T. Nagel (1985) Agent-relative morality, also in *The Ethics of War and Nuclear Deterrence,* pp. 15–22.

[12]In consumption, when we spend on guns we have less to spend on butter. Suppose that the guns are more or less worthless, but that spending the money on butter would not have been. We cannot argue that buying guns makes little difference because much of the money we spend helps others by employing them; that is *also* true if we had spent the money on butter, *and* we would have gained butter's consumption value as well.

[13]The general argument pursued here is very similar to Norman Daniels' "Prudential Lifespan" Account. *See* N. Daniels (1988) *Am I My Parents' Keeper?* Oxford University Press, New York, NY, pp. 40–65, 83–102.

[14]Much of the substance of this chapter is developed at greater length or in a different way in Menzel, *Strong Medicine* (*see* Note 1), ch. 3, 4, and 11. Here I have assembled things in quite a different way, and in the second and third sections ("What Does It Really Cost To Save a Life?" and "Ought We To Count All the Real Costs?") of the current chapter I have substantively altered some conclusions in chapter 4 of *Strong Medicine.*

Intent and Actuality

Sacrificing the Old and Other Health Care Goals

Jane A. Boyajian

This chapter illuminates factors that shape health care goals affecting the old. I explicitly avoid the obvious: the implications of costs and demographics are addressed elsewhere. Rather, this is a discussion of the less obvious. It demands a new standard for self-scrutiny and proposes new tools for goal assessment.

Health policy emerges as a win, loss, or compromise in the real world of the biennial budget, the legislative caucus, gubernatorial priorities, agency realities, and citizen input. Stated values are visible in the goals of statutory preambles, state of the state addresses, and institutional mission statements. However, another less apparent dimension informs the political process for, just below the surface, other goals and values also shape policy formulation. I intend to explore this dimension and the values embedded in it.

Public policy is the ground in which the values that people articulate are actualized or not. Because state laws and administrative policies require compromises between competing interests, reality does not always support stated intent. Thus, there is often a vast difference between the goals we articulate and the values actualized in our policies regarding the old. In a sense then, this is a chapter about gaps—the gap between theory and practice, the stated and the unstated, values as intent, and as actuality. I begin by iden-

Aging and Ethics Ed.: N. Jecker

tifying several current goals; although these appear to be competitive, policy formulation should actually strike a balance between them. I then review several policy questions about the old. In the course of this review, I mean to illuminate the interiority of these policies by asking: What values do these policies actually reveal? Finally, I offer several principles that I believe should shape the way we think about care of the old. As we formulate policy and assess goals for the future, these principles add a markedly different dimension to ethical analysis.

The Current Environment

Vigorous change shapes the environment in which health care goals are now being advocated and implemented. Two themes, in particular, will likely dominate the discussion in the 1990s—cost containment and access. Professional association meetings, learned journals, and religious council meetings are widely involved in considering what state and federal changes are needed to address these concerns.

Some regard the question, "Access to what?" as one primarily about rationing resources.[1] For example, in 1989, declaring that a state of emergency existed, the state of Oregon established a uniform benefits program to provide health care to all persons under specific income levels. A Health Services Commission now must prioritize the services the state will provide and develop a method of reducing these priorities in case of a revenue shortfall.[2] Only those services are to be offered for which there is sufficient funding; services ranked below that funding level will not be available.

Some applaud Oregon's chosen course to make health and medical rationing goals explicit: "We're rationing health care already...When we set the budget each year, we somehow escape responsibility... Isn't it better to say it out loud, under public scrutiny, that these are the decisions we are making?"[3] Others are appalled that the rationing affects only the Medicaid eligible: "The very formulation of such a list is 'outrageous'...What's awful is that...the state of Oregon...[is] only putting the burden of health rationing on the poor."[4]

Still others believe the issue is a matter of equity and the question should be formed differently to ask instead: "What care shall we provide all citizens?" Thus, whereas the possibility of rationing is an undercurrent in Washington state's discussion, public and private sector spokespersons today place the focus on the provision of basic health to all citizens. So the question becomes, "What basic health services should be available to all the state's citizens, regardless of ability to pay?" In a sense, Oregon wonders: "What ceiling shall there be?" Washington, though, asks, as did one Presidential Commission: "What floor shall undergird everyone?"[5]

Now the reader should not assume that statewide consensus exists either in Oregon or Washington. However, even at this early date in the discussions, these two approaches suggest a different weighing of values that transcend mere semantics. Although each state may ultimately offer (or restrict) similar services, the values that appear to shape the discussion diverge. What is most exciting about both approaches, however, is the urgency and seriousness with which health care reform in these two states is now being considered.

The Will to Change

Of course, no guarantee exists that these efforts will readily result in a major re-formation of medical and health services financing and delivery, or in effective strategies for addressing access and cost containment. However, several trends indicate a deepened interest in health care goal-setting and point to a unique climate for change in the United States. First, the public's dissatisfaction in our health and medical delivery systems is becoming more evident to decision makers.[6] Second, state governments seem increasingly reluctant to await fiscal, program, or moral leadership from the Federal government.[7] Third, there is an emerging recognition that our funding to date has promoted a *medical* care system (crisis intervention), whereas a *health* care system would, instead, give priority to wellness care and disability prevention. Fourth, the long-term yield of health promotion and disability prevention strategies now seems both fiscally prudent and morally necessary to corporate purchasers, providers, legislators, and the constituents they represent. Fifth, more federal and state leaders recommend that long-term care become a new entitlement than ever before. Thus, a will to change is emerging across sectors, promoting alliances between purchasers and providers where they have not existed before.

The Washington State Medical Association (WSMA) succinctly captures this spirit in a House of Delegates resolution calling for a basic health plan universally available by the year 2000. It declares that:

> ...there is too little agreement on what services are appropriate or what constitutes quality...[Whereas] the health care "system" is accessible, offers high quality and is affordable for many Washingtonians...[f]or many others, there is no "system" at all...Today nearly 785,000 (17%) of our state's citizens lack any health insurance...[yet the] health care system in our country has been predicated on the principle of universal access—that all Americans, regardless of income, should have access to the system and its cornucopia of services...Further study is required to determine appropriate consumer participation and funding mechanisms for these desired services for the population.[8]

Moreover, the WSMA House of Delegates ..."actively encourage[s] a public dialogue to form a consensus as to what constitutes basic health care services" in Washington state. It is the spirit of this resolution that gives hope, for it captures this new will to change.

Political Realities

These resolutions are a welcome challenge to a phrase used in the public sector—"the elephant's too big." That phrase sums up both the complexity and scope that significant reform would entail. Moreover, there are several inescapable realities with which reformers (policy analysts, citizen advocates, agency spokespersons, and legislators) must contend. Theoreticians who suggest policy strategies without taking into account these realities are of little help to public sector decision makers. Conversely, policy makers who permit these considerations alone to drive their strategies fail to serve the public's interest. These realities include the following.

Piecemeal Approaches

Especially in health and medical care reform, meaningful analysis requires comprehensive assessments of many interdependent issues. The sheer complexity and scope of health care reform defy facile discussion and promote a piecemeal approach; hence, long-term care problems usually are divided along categorical lines as if the concerns of the old, the developmentally disabled, veterans, and the mentally ill were totally separate considerations. Agencies reflect their mission. Health, ecology, and human services, for example, are examined as separate rather than interdependent worlds. Other realities promote bite-sizing: most legislatures are part-time, and their sessions are necessarily compact; the volume of work compressed into long legislative days requires succinct analyses that blur subtle distinctions.

Comprehensive analysis is difficult in arenas that are reliant on demonstrating short-term impacts. For example, both federal and state fiscal and program planning are normally taken in two-year bites. So what cannot be accomplished within a biennium may not be feasible: major reform could cost legislators' constituents, and the benefits may not be apparent until several elections hence, when the legislators who risked the reform are no longer in office. In addition, the public's or media's attention to issues (which energizes the legislature) is usually short-lived.

Feasibility

It is no surprise to public sector veterans that the criteria used to determine goals in health care reform include several additional factors beyond those that assess the goals' inherent worth. Let the following example make explicit factors that often circumscribe the selection of public policy goals. Asked to recommend criteria for prioritizing health goals in one state, a task force first readily identified these five: impacts on health and disability, cost-effectiveness analysis, economic risks and benefits, the legal authority to act, and the technological capacity to address the goal. All these considerations are obvious. However, task-force members were not only knowledgeable about health care goals, but also politically astute. Not surprisingly, they made explicit additional criteria for selecting health goals that took the political environment into account, that is whether they could be actually achieved. So they asked: What are the trends? This criterion evaluated the degree of constituent interest as well as the technological state of the art. How aware is the public? Is a crisis emerging that will capture people's concerns? Statewideness was added to the list; that is, do the majority of the state's communities regard the goal as relevant to their interests? Can consensus be developed easily? Finally, can significant movement be made in the next biennium? Above all, the task force noted the importance of a strong moral concern in the community by asking: Is there a public sense of outrage? Is it sufficient to create a will to change and make that evident to the legislature and the governor?

In a subsequent draft, the following were enumerated as criteria for selecting the state's priority health goals:

1. Would efforts in this area be relevant to the majority of communities in the state (even though the specific related needs, objectives, programs, and resources might vary among communities)?
2. Can *some observable progress* related to the goal be made by the end of the biennium?
3. Would substantial benefits to human health be likely to result from programmatic efforts related to *objectives* in this area?

4. Would substantial risk to human health be likely to result from a failure to give priority to programs related to this goal during the next biennium?
5. Is there an opportunity to make progress in this area during the biennium that might be lost if we were to wait "until next time"?
6. How readily can consensus concerning needs, objectives, and programs related to this goal be developed, e.g., between: those who govern and those who are governed (politicians and constituents); the executive and legislative branches of state government; political parties; House and Senate; state and local governments; various special-interest groups; responsible executive agencies?
7. Would substantial economic benefit result from priority attention to programs in this area?
8. Would substantial economic cost result from a failure to give priority attention to programs in this area?
9. Are legislators already at least aware of the need to give attention to this goal (or, would it take the whole biennium to educate them)?
10. Is there any kind of "trouble brewing on the horizon" related to this goal, that is, is there any evidence that the state of Washington might face one or more crises related to this goal during the biennium?
11. Does the state have the authority required for spear-heading action related to this goal, or at least, can we expect to be able to do *something* without encountering legal barriers?[9]

These criteria reveal political factors that are usually hidden from view. Most interesting is the sheer weight of the criteria evaluating feasibility, that is, whether or not a goal is realistically achievable.

My own inclination is that responsible policy decision making should do just this: make these implicit factors apparent and open to scrutiny and, therefore, accountability. Furthermore, although political factors should not be the sole determinant in choice making, failure to consider them is as irresponsible as selecting health goals without assessing impacts.

Balancing Interests

In health reform, several competing tensions now shape goal-setting. Although changing citizen interest and legislative or gubernatorial priorities may generate emphasis on one over the other, these are best understood as competing values or interests that pull in opposite directions. In reality, I believe these goals must be balanced; they should be in dialectic. Among the polarities that must be balanced, the following are especially critical.

Balancing Cost and Autonomy

In discussions about the old, this tension is usually presented as the polarity between elders *demanding* access to whatever interventions they choose (regardless of cost, utility, or outcome) and the young whose access to needed resources is restricted by elders' excess. I put this issue aside, for the moment, to note that self-determination and cost containment both are values with which a responsible society should be concerned.

In this democratic society, personal autonomy is understood, in part, as the right to control the circumstances of one's care. Privacy, in this context, means the right to refuse or accept services based on one's own values. At the same time, cost containment and resource allocation have become urgent matters as we begin to understand that choices about medical care affect the commons and may burden the future.[10]

The consequence of concern both for cost containment and autonomy is that public policy goals now promote home care over institutional care, since both cost reduction and personal autonomy are served by home care. Home care is the preference of the old; fear of institutionalization, with the result of *losing control* over one's life, is the greatest fear to citizens polled about long-term care and dying. Also, for the moment at least, the cheapest venue for providing care is the home. Unfortunately, trends suggest these two goals are on a collision course. Home care may become the more costly alternative as disability and frailty increase with an aging population, DRGs promote early discharge of the very sick old, and the costs of providing home services in rural areas grow. When home care becomes more costly than congregate care alternatives, which goal will we value more highly—autonomy or cost containment?

Balancing Decentralization and Consumer Protection

Global vs local control is at issue here. Several examples illustrate what is at stake. Currently, deinstitutionalization and the expansion of community-based care settings are both federal and state goals. At its best, deinstitutionalization should promote patient independence, improve personal control over one's care, and generate the community's sense of ownership for services. Moreover, decentralization promotes another goal important to constituents—regional relevancy, meaning that health and social services should reflect regional values and needs. All these considerations argue strongly for decentralization to more relaxed care settings.

However, balanced against these important goals are several others we also value. For example, how do we promote regional flexibility while assuring that services throughout the state are comparable (statewideness) as required by federal programs and valued in a society committed to equity? Moreover, statewideness constrains innovation tailored to specific consumer needs. Distances in some regions are so vast they preclude cost-effective ways to provide a continuum of needed services. Furthermore, quality assurance and consumer protection are difficult goals to achieve in decentralized care settings that are designed to be informal and home-like. Even so, new federal mandates now require establishment of consumer hotlines for home care clients, and state requirements are now more likely to mandate licensure of hospice and home health care agencies than before. These tools address the goal of consumer protection, but intrude on the home-like atmosphere that the goal of decentralization and deinstitutionalization promote. The question is: How do we keep these goals in balance? How do we titrate the goals of deinstitutionalization and consumer protection?

Even agency mission and size, which are set by legislatures, reflect the pull between decentralization and consumer protection. When the public's perception is that agencies are too large (meaning becoming bureaucratic and unresponsive to the consumer), legislatures break programs apart creating smaller agencies or privatizing programs to promote responsiveness to consumers. Such splintering, however, complicates coordinating services for citizens whose needs do not divide easily along legislatively determined categories. When, however, improved service coordination is a higher goal, legislatures consolidate programs into conglomerates.

Balancing Generic Services and Categorical Case Management

Should a citizen's access to services be through a single point of entry (generic) or directly to the specific programs a citizen needs (categorical)? These goals shape the organization of the services that individuals seek. The problem is extraordinarily complex, and interest groups lobby in both directions. Today, citizens enter social and health services through categorical programs, for example, through mental health, drug abuse, or home care. So specific symptoms and needs define (categorize) the client. When service delivery is organized categorically, it is difficult to address a client's total needs; drug abuse and Alzheimer's disease are treated as if they were totally independent issues. Thus, several caseworkers may be assigned to a client, each of whom works with a single category (e.g., mental illness, alcoholism, home care); this creates case coordination and integration problems. A single point of entry provides a central office to which individuals (no matter what their needs) are referred for assessment. However, because there is such a broad range of services available, each with differing eligibility requirements, one generic case manager, working alone, may have difficulty keeping abreast. In adding staff to address this complexity, the initial access point then becomes simply another referral point.

Two trends exacerbate the polarity of these goals. On the one hand, advocates for the developmentally disabled and the mentally ill argue strongly for continuing categorical designations, because they fear generalists will devalue their clients or cannot be responsive to categorical needs. On the other hand, among the aging, for example, the population of shared or crossover clients who fit several categories simultaneously is growing, so complicating coordination. For example, a client may be old (and needing services from the aging network), a substance abuser (thus requiring access to drug and alcohol abuse programs), and mentally ill (programs for whom are shaped by very different state and federal mandates). An elder may be a resident in a nursing home managed by the Department of Veterans' Affairs while also requiring services through the Department of Health and Department of Human Services. A prisoner under the jurisdiction of the Department of Corrections may be a drug addict and an aging victim of Alzheimers. An aging population suggests that the crossover client will be more the rule than the exception in the future.

Balancing Fiscal Integrity and Client Service

Whether there is a single point of entry or a categorical one, people need knowledgeable service coordination so they do not fall between program areas. Good coordination requires cooperation between many public and private agencies. However, in a declining economy, reduced funding generates competition, promoting agency aloofness to preserve the distinctive mission and *product* that garner foundation, legislative, and public support. While cooperation is imperative to serve clients' interests, agency survival feeds competition for clients and a market share. After all, the ability to show a profit and vigorous client use is a survival tactic. So the drive for survival creates conflicts of interest for staff, since what is in the agency's interests may work against a client's. Client needs can compete with profitability. The agency may continue services to a client longer than appropriate; an agency may inappropriately evaluate a client's need for its services. Conversely, agencies may create barriers to citizens, especially those with complex or undesirable problems, thus keeping client loads and agency costs down. The ethical ramifications of these conflicts between client and agency interests are now a subject of concern for the involved professions.[11]

Summary

Several important competing goals have not been addressed here because the literature provides opportunities to reflect on these: achieving a balance between health promotion and acute care, and between cost containment and quality of care. Self-determination and protection constitute another polarity significant in any discussion about aging since we have both a responsibility to promote the individual autonomy of vulnerable people while also protecting them from harm.[12] However, let the examples addressed above illustrate my point: health policy decisions involve a difficult balancing of worthy competing goals. Just as ethical discourse involves reflecting on competing claims, deciding between competing health care goals is the work of state government. The hope is that a balance can be found that serves justice and compassion and that is, at the same time, politically realizable.

Examining Health Care Goals for the Elderly

In examining the values reflected in the health care goals we choose, we should ask two questions: First, what values do we intend to promote when we select one goal over another? Second, what are the results of chosen policies?

This inquiry enables us to identify the gap between possibility and reality—between intent and actuality. It is the second question, however, that interests me as we review several policies with regard to the old. Uncovering this dimension also enables us to discover instances when our own self-interest obscures our stated intention, and we are caught in moral quicksand.

In exploring policies, we discover what values actually drive them. One study explored this dimension and identified four values that have shaped our health and medical policies: professional autonomy, individual autonomy, advocacy, and access. It noted that quality is a more recently added value.[13] My own belief, however, is that professional and personal autonomy determines our policies in practice, and that advocacy is honored only when it does not encroach on the autonomy of the empowered. In reviewing two policies particularly as they affect the old, my intention, as before, is to illuminate what is usually hidden from view. In examining these policies, we ask: What values are *actually* affirmed?

Caregivers for the Elderly: Stinginess as Social Policy

Promoting long-term care in the home, rather than in institutions, is the stated intention of most public and private agencies. After all, home care affirms values we espouse: personal autonomy (since most elders prefer to remain at home); the least restrictive alternative (so intruding on individual liberty as little as possible); and stewardship of resources (since home care is cheap, for the present). This objective then follows: expanding community-based services to support patients and their family caregivers and maximizing the functional independence of elders. Although our stated goal is promoting home care, there is a chasm between intent and actuality into which the families and friends who care for their elders fall, making family caregiving a dangerous undertaking.

Caregiving Realities

Most elders age in place and care for themselves as independent people. However, many elders require assistance sustained over long periods of time, and demand great physical and emotional endurance. The Commonwealth Fund Commission on Long Term Care Assistance found that "in excess of 5 million of the 29 million elderly people living at home suffer from functional limitations that restrict their independence. Within this population, 1.6 million elderly people are severely impaired, and 300,000 live alone."[14]

Of the 1.6 million needing long-term care assistance, 1.1 million have severe disabilities requiring assistance in activities of daily living (eating, personal care, lifting), and an additional 300,000 have severe cognitive impairments making further demands on their caregivers.[15] Impaired elderly are more likely to experience chronic disease, and generally are older, poorer, and more dependent than other elders. Whereas formal care providers assist 3% of this population, 70% of the care is provided primarily by family and friends, more than half of whom are wives and daughters. Up to 30% of our workforce now provides elder care.[16] Projections that the population 85 years or older will increase by 46% in the next decade should cause additional concern: the care needs of these old-old are far greater than the young-old; the old-old are predominantly women who are usually both poor and single.

Just as the old-old tend to be women without resources, so too their informal caregivers generally are poor and lack resources. Briar and Ryan, having reviewed the negative impacts of caregiving on women, note that the policy goal that promotes aging in place has profound consequences on women caregivers that have yet to be addressed.[17] For example, "...of the 2.2 million people who cared for the 1.2 million frail elderly in 1982, 71.5% were women. While less than 10% of the caregivers quit their jobs to take on the burden of care, a sizeable proportion had to rearrange their schedules, reduce their work hours and take time off without pay."[18] Moreover, nearly 90% of caregiving families have modest means, whereas the income of 30% is less than 125% of the poverty level.[19] With the aging of the old, it is also true that these caregivers are themselves aging and often affected by their own disabilities.

The Costs

The cost of family caregiving is emotional, financial, physical, and social. Given what we know about the feminization of poverty, the personal toll for family caregiving is especially harsh for women. Here is the poignant reality for most caregivers: having undertaken caregiving, they have little awareness of costs to themselves. Although society relies on their caregiving, we fail to support them, regarding them, instead, as free laborers. Briar and Ryan speculate that the reasons women accept this caretaking function so readily are because they: expect to provide that care without support (having been trained through social norm), applaud the trend toward deinstitutionalization, and are not accustomed to demanding that legislatures and state agencies support their efforts.

There is a decidedly dark side to this situation. In the last decade, we have noted a rise in the incidence of elder abuse and neglect. Certainly, many instances are perpetrated by strangers or by family members who have learned family violence or are substance abusers. However, the fact remains that much abuse and neglect are caused by family caregivers who undertook caregiving with good intentions, but they endured its rigors unsupported since we do not properly fund services to support caregiving at home or to assist care providers themselves. They become exhausted and stretched beyond endurance, for studies show that the average abused or neglected elder is over the age of 80, has at least one major disability, and has lived with a care provider for more than ten years. Thus, well-intentioned caregivers simply implode during the caregiving process.[20]

There is a misogynist aspect to this situation; since most family caregivers are women, we should not be surprised that the task of caregiving is trivialized. However, a just and compassionate society must "...direct [its] explicit attention to women caregivers both as client and as service providers and...explore the extent to which the premises underlying deinstitutionalization and diversion [community care placement] are *de facto* sexist."[21] Callahan reminds us of the "heroic self-sacrifice" that caregiving entails. The vulnerability of this self-giving requires something in return; he says that caregivers "...cannot themselves be left in self-sacrificial isolation. Our burdens must be pooled and shared."[22] Yet, in the

United States, our public policies fail to fulfill our stated goal of supporting home care and caregivers' labor. They are left in devastating isolation, illustrating the enormity of the gap between our intended goals and reality.

Resources for the Elderly: Sacrificing the Old

Contrasting Views

Two antithetical perceptions of the old bias our discussions about their impact on the medical commons. One view incorporates a picture of an empowered political block with financial means, whose growing numbers combine with the cost of their entitlements to threaten the well-being of other groups. This view holds that these old "...are already costing too much and in the future will pose an unsustainable burden on the American economy."[23] Proponents of this view hold that the unrealistic expectations of society and of physicians that death can (and should) be held at bay compound the burden. One perception is, then, of an entirely self-centered, self-serving group whose unwillingness to accept death's inevitability denies others.

The second perception is of an aged population that has become fearful of the end of life, not because death will come, but because death will not be permitted. These elders believe that, because of others' commitments to teaching and research goals or fear of litigation and moral slippery slopes, they will not be allowed to die naturally. This contrasting view holds that our aging population is preoccupied with the concern that they will not die soon enough, and that they will lose control over their lives and their deaths in medical settings.[24] Although there is some evidence supporting the former view, polls support this latter perception.

Age Rationing

One question dominates the discussion today: how to balance the use of medical resources by the old (for whatever reason) against the limited resources available and the needs of other worthy groups.

Instead of examining causes, we discuss symptoms.[25] I believe that we should be asking what factors actually drive costs and the choice to use expensive interventions. Even Callahan (the most discussed proponent of the former view) states that "the evidence is good that age by itself does not drive up health care costs, but that age combined with the application of expensive technological medicine and improved services does."[26] Of the alternatives proposed to distribute resources more equitably, most are rationing schemes. Indeed, one analyst lists these approaches to rationing: age; ability to pay; residence; entitlement; need; effectiveness; attractiveness to screeners; acceptability as research candidate; lottery; first come, first served; social or moral worth; power of specialty or advocacy group; health, legal, or economic risks without intervention; political, media, or public pressure.[27]

Categorical rationing (or "disenfranchising the old" as Jecker labels it) is prominent, in part, because it is clear-cut.[28] Indeed, one proponent for age rationing claims its advantage is the ease with which people can be classified. Others advance age rationing as a matter of principle (social justice), whereas another proponent claims that, even if resources were not scarce, age rationing in itself is a benefit to the old.[29] I contend that age rationing asks that elders sacrifice themselves for other reasons; it is easy: it supports pervasive biases of the dominant culture; it saves us from having to restructure the biomedical delivery system.

Sacrifice as Duty

First, age rationing is easy because it is an easy concept to sell to the old. After all, parents will usually sacrifice themselves so that their offspring might survive. By portraying the old as obstructionists to the future, proponents of categorical rationing have fired a cheap shot. Proponents of rationing present a view of old age as synonymous with consumption (taking from babies), self-centered (unwilling to deny self for other), unproductive (disabled and ill), useless (whose unproductivity is released by death), and dependent (child-like). It is easy to ask people so defined (and who come to believe this of themselves) to give up their place, especially when they also hold the view that no sacrifice is too great for the future.

Indeed, generally speaking, elders themselves fear that their prolonged dying or chronic illness will become an obstacle for their children; either their economic or social dependency is thereby created and/or the resources the parent intended to leave to her/his children must be depleted for her/his care. These advocates of age rationing have played so successfully on this natural parental concern (buttressing their proposal with statistics about the high expenditures in the last year of life, which we will address later) that it is not unusual to hear elders now state that they should remove themselves "to make way for the young."

Selling age rationing sometimes invites revisiting this culture's romanticized view of Inuit elders. In the 1970s, as we struggled to break through the death-denial of our culture, the vision of the Inuit elder dispatching himself/herself to the ice floe to die was paternalistically presented as the *noble savage,* more wise and accepting of death than we. In the telling of the tale, we overlooked the reality: Inuit did not easily remove themselves to the ice floe; they did so reluctantly and fearfully because *there was no other option.* The winter having outlasted the food supply, in that context the elder's death was the only way the rest of family had a chance for survival.

Although our circumstances today require major social re-formation, *we are scarcely in extremis.* Recalling Childress' discussion of conflict situations when all cannot be saved, we are simply not at the point when many must destroy themselves so that others can be saved. This is not the time for "morals of last days," for no "apocalyptic crisis" exists that "if none sacrifice themselves of free will to spare others—[we] must all wait and die together."[30] How can we then be satisfied with a system that calls for the self-sacrifice of a vulnerable group (a group predisposed to sacrifice because they are the parent generation and because they are predominantly women socialized to first give up self in behalf of other) *before* other options are explored? I do not agree with others that we have only two choices—categorical rationing or case-by-case decisions.[31] There is another option that, if addressed, can more equitably meet the problem of scarce resources *while also* addressing basic human need. That option is to re-form the biomedical system so that it serves other values than the professional and personal autonomy of the privileged. This is an option that asks the empowered to give up something.

The Old as Evil

Second, age rationing is easy because old age has become a new metaphor for evil. Thus, age rationing reflects our biases; by consigning the old as other, we have solved a problem of evil in our time. Susan Sontag offers perspective here: "The melodramatics of the disease metaphor in modern political discourse assume a punitive notion: of the disease not as a punishment but as a sign of evil, something to be punished."[32]

By making the old *other,* we can more easily dismiss them and our commitments to them because they are marginalized, at the periphery of our worlds. There, they seem incongruous with life, ambiguous persons—says Ronald Philip Preston. He reminds us of Gregor's gradual change into a cockroach in Kafka's *The Metamorphosis;* as he changes so does his family's perception and treatment of him. The family recoils from him and is angered by his presence. Over time, his family blames him for his fate (and theirs) and gradually ceases to regard him as of them. He becomes other, so his death is a happy event for the family because it is released.[33]

Preserving the Status Quo

Third, age rationing is easy because it supports the status quo. If categorical rationing is successful, other factors that drive health and medical costs and that deplete the medical commons need not be addressed.[34] One analyst comments that:

> ...the inappropriate deployment of medical interventions *insofar as it occurs,* is not impelled by the demands of the elderly or

> their families so much as by professionals setting the wrong objectives or working under extraneous and unnecessary pressures from the administrative arrangements for funding or the fear of litigation.[35]

A strong case can be made that both the imperative of technology (once we have it, we must use it) and imperative of academic medicine (which promotes research), rather than age, drive health and medical costs. Addressing these drivers would force us to an undertaking that would affect all those of us who take access to health and medical services as a given in our lives. We would need to undertake a major restructuring of research funding and determine what services should be offered. Clinical disciplines would thus be reformulated. Such a re-formation would ultimately impinge on the very values that have driven medical and health goals to date (as noted earlier, these being personal and professional autonomy).[36]

In summary, limiting elders' access to medical resources because we have decided that they are the cause of our resource shortage is a diversionary tactic. Scapegoating, says Binstock, has a salutary effect for decision makers and care providers, since it diverts

> ...our attention from a variety of deficiencies in political leadership and public policy...engendering intergenerational conflict...diverting our attention from longstanding issues of reform involving policies that provide benefits to older persons.[37]

The tactic of blaming victims for their ill health when they cannot control the social, commercial, and economic causes permits retrenchment. Here, scapegoating the old distracts us from the socioeconomic realities that promote ill health, disease, and dependency.[38] Moreover, we are diverted from the factors that drive the development and use of high-cost technologies and promote acute care services instead of expanding the availability of basic health services.

In summary, age rationing is offered as a strategy for achieving a more just distribution of the medical commons. Instead, it places particular burdens on women, the poor, and the disadvantaged while protecting a biased system from reform.

Principles That Matter

We have explored competing health care goals that, in practice, require balancing and the impacts of two policies on the old, highlighting the gap between intent and actuality. We claim that we support aging in place, reserving institutional care only for the most needy; in reality, we support a policy of neglect since we are unwilling to finance services that support living at home. This neglect has especially harsh consequences on single old women. Age rationing, if adopted, is a policy of sacrifice, asking the poor old to give up their claims on the medical commons for the sake of others (or, as Callahan claims, for their own sake). Both policies are morally suspect because they (1) serve the interests of the empowered, asking only vulnerable people to give way, and (2) fall most harshly on the disadvantaged old. A just and compassionate society should explicitly push hard against the tendency to protect the status quo because it perpetuates the burdens of vulnerable people. Instead, policy-makers and theorists have an obligation to be conscientiously critical of their own intentions (especially those hidden from view) and of the policy goals they espouse (especially their impacts on disadvantaged people). This evaluation should include several principles beyond the customary. The purpose of this closing discussion is to identify these principles and their significance to vulnerable people.

Honor Special Covenants

A covenant implies a commitment in which the entrusting

> ...of one party to the other or of both to each other...obligates them to stand accountable. ...Thus acts of entrusting bring covenants into being. ...[Those who enter into covenants] affirm

> (implicitly or explicitly) that they belong to the same moral community...meaning [they] recognize one another's worth, and not merely their usefulness.[39]

Beyond the covenants between two individuals or between humanity and God, Joseph Allen describes a third, the special covenant. This covenant is "...an intermediate level of human moral relationship, one in which we are related to some people in continuing ways that are morally significant."[40]

My belief is that *we have (and ought to honor) a special covenant with the vulnerable and disempowered.* That special relationship exists, in part, because their vulnerability is primarily caused by our failure to address the economic, political, and structural conditions that disempower people; our very own empowerment often comes at their expense.[41] Our covenant with disempowered people is of still greater significance when they are old, since the old have moral claims on us also because they are our parents. They nurtured us individually when we ourselves were vulnerable and, as a society, we have a special relationship with them because their collective yesterdays created our presents and futures. Furthermore, we have affirmed that covenant by promises we have made in public policies on which the old have relied; for example, Social Security is a covenant between generations. Any changes in those policy-promises must be critically assessed not only on their own merits, but also because the old have made life plans trusting that we will honor our commitments.

Does it mean that commitments can never be broken? Discussing the circumstances when obligations might be broken, Margaret Farley stresses the great cost of breaking them: "promise-keeping is one of the foundations of society. When it is commonly expected that promises will not be kept, people lose their moorings; society staggers into chaos, often, on the brink of violence."[42]

However, since some commitments compete for our allegiance and, since we cannot honor every claim, we must make choices with the result that we may break commitments. Sometimes a commitment requires that we sacrifice ourselves in behalf of our obligation, but as Farley says, there are limits to self-sacrifice *especially* when there is an imbalance of power: "When a disproportionate burden of sacrifice is laid on one person in a commitment-relationship, and when the person who bears it is the one with the least power, the duty of self-sacrifice is morally suspect."[43] Surely we can see that health and medical policies that place the burden of sacrifice on the old are morally suspect.

So policies that break commitments with the old have profound consequences: First, we have broken a trust and, in the case of elders who relied on our commitments embodied in Medicare and Medicaid, for example, the breach that results cannot easily be filled. Second, we have hurt ourselves, for our confidence and trust in what is promised us in public policy is eroded. Thus, breaking commitments with the old is morally perilous and should be considered *only* when no other options exist.

Reach for Size

A responsible person reaches for size. By that I mean reaching beyond piecemeal approaches is a moral responsibility.[44] A policy of size is one that escapes (insofar as possible) the limitations of one's bias and community. It reaches for large possibilities or what hope theologians call utopian dreams. These are realizable dreams because they are grounded in accurate assessments, yet imagine a future in which people can live more fully. We have an obligation to reach beyond the obvious to possibilities of great size, which are more inclusive rather than less, and to push beyond self-interest.

A health or medical policy of size is one that promotes the well-being of all over the continuing privilege of the few. Imagining a future that is simply a variance of the past-present and so promotes our advantage is morally suspect. Such is the case with the two policies discussed above, since neither ask how we can construct a future that includes those outside the system. Reaching for size demands that we imagine (and undertake) a radical reformation of our present health/medical models.

Avoid Bias

We have an obvious obligation to ensure the accuracy of our data. We should subject our interpretations of data to critiques so as to purge them of our biases. This critical analysis is obligatory when we use the data to buttress proposals of such magnitude that we consider suspending our commitments to a vulnerable group. When imprecise data perpetuates bias and inequity, it is a moral problem. Unfortunately, policy debates about health and medical goals are subject to this imprecision.

Most importantly, our terms often are imprecise: we do not explicitly acknowledge that the old of whom we are speaking are principally poor women, so we mask our misogyny. In discussing resource allocation and age rationing, *medical care* and *health care* are terms used interchangeably, yet they involve different goals, means, and professionals. Our cost estimates and outcomes mean little if we cannot distinguish among:

1. promoting well-being;
2. supporting functional independence;
3. addressing chronic care needs;
4. responding to crisis; or
5. postponing dying.

When lumped together, we cannot evaluate the cost-worthiness of any intervention.

The price of providing care to elders needs clarification. All elders are neither disabled nor institutionalized, since only 5% of those over 65 are in nursing homes, and only 23% of those over 85. Most elders are independent or receiving care from friends and family. In 1987, of the $500 billion spent on health care, more than 50% financed hospital and nursing care; given these settings, we should assume that expenditures were for illness care for a small percentage rather than for wellness care and disability prevention.[45] To imply that most elders drive these costs is inaccurate and promotes scapegoating.

Statistics on disability are misleading because disability does not automatically equate with a reduced quality of life (whatever that means). For example, although 69% of disabled men over 65 reported a disability, their working level remained constant.[46] We cannot even be sure what statistics showing increased disability among the old mean: Is disability among the old in itself increasing, or do individuals better evaluate their disability (a measure of effective education) and report it?

Promote Citizen Decision Making

The town meeting approach to decision making is now reemerging as a consensus-building tool in bioethics.[47] The effectiveness of these statewide projects to promote citizen education and choicemaking has been mixed. The Oregon model has been effective: since its inception in 1983, succeeding forums for Oregon Health Decisions have forged a public consensus that change was needed, opening the way to Oregon's 1989 rationing plan. However, even at their best, these public forums tend to involve individuals who are accustomed to public advocacy rather than silent voters or marginalized persons. In Washington, the legislature recognizes the importance of public involvement and so mandates public forums on community protection, health access, cost containment, long-term care, and the like, yet policymakers also recognize that these forums can become the captive of lobbyists and single-interest advocates.

Lean Toward the Disempowered

Policy options we are now considering could result in renegotiating covenants with vulnerable groups; so finding effective ways for their involvement should be regarded as a moral responsibility. Individual realities differ, so we are obligated to examine how policies affect individual lives. Attention to the consequences for vulnerable groups is required, since they are so often negatively affected. Individual realities should mediate theory as a tool for making the hidden dimension visible, and the gap between intent and actuality clear. This responsibility has greater significance with regard to disempowered persons whose placement outside decision-making processes is complicated by their distrust of established systems.

Decision makers must be affectable, meaning that they are obliged to be changed by the disempowered. We "...have an obligation to open ourselves to others so that we are fundamentally changed by what they know and who they are."[48] Our openness to be changed by the realities of disempowered people gives them the power to influence, and assists us in resisting our tendency to homogenize, blurring the distinctions between people. Personal histories assist empowered people's abilities to hear and be affected by that which they cannot know—how individuals experience ageism, racism, and sexism.

Address Long-Term Impacts

A truly restructured medical and health system whose goal is a more equitable system, and that stresses wellness care over acute care, will have consequences for the medical establishment. Clinical education has emerged from an acute care model rather than one giving priority to preserving health or improving the functional abilities of disabled people. A shift to wellness care will require academic medicine and the teaching hospital to change radically and renegotiate their covenant with society.[49] There are two issues here. First, will academic medicine recognize its responsibilities to serve the public, rather than serving its own research interests in the name of academic freedom? The public has so generously funded academic medicine that it should be regarded as a *public trust* carrying with it an obligation to address needs that the community has identified. Doing so would ask academic researchers to forgo their absolute autonomy in selecting research goals in order to undertake "...several important social omissions including education, patient care, and research toward improving the health of the public."[50] Clinicians would then conscientiously rethink "the rule of rescue...[as] a deontological imperative" as Jonsen describes it; that is, they would reconsider their inclination to attempt to rescue every "doomed" patient.[51] If these shifts in health care goals were actualized by academic researchers and clinicians, then the community as a whole would itself incur an obligation to address the resultant structural dislocation in the medical system. Responsible planning would involve strategies for addressing the negative impacts on specialty areas, medical schools, and hospitals that are now reliant on providing acute care research and services.

Summary

A just society should reach beyond self-servingness. A compassionate society will take particular care to lean toward those whom the status quo denies—the old, the disadvantaged, and the young. They need not be adversaries. We should remember that one's value should not be measured in one's willingness to sacrifice self so that others may continue to enjoy a full range of services, or their research or clinical goals. Rather, the personal and professional autonomy of some (however well disguised) should cease to be the primary value that drives our health and medical delivery systems.

We should, instead, be concerned most about promoting mutuality, and building a society in which the empowered reach beyond themselves. Such a society will value advocacy for those whose quality of life is limited. Rather than concentrating on the limits of care to the poor, we would be considering what basic health and medical services will be available to all, and what changes we (the empowered) must initiate so that that goal is attained. We should reach for size, for, as Kapp says, we have "...an obligation to suggest and help implement social policies...[that] do a better job of promoting social harmony, protecting the legal liberties and entitlements of individuals, and honoring the ethical precepts of autonomy, beneficence, and distributive justice."[52]

The options before us need not disrupt our delicate social balance. Rather, we can reach toward size, a vision that recognizes the import of giving up some personal power so that more can have access. Theorists have an important role here. As Ernst Bloch has said: "philosophy...especially ought to bear the torch before and not the train behind."[53] It is an important effort and calls for our best energies and our highest commitments.

Notes and References

[1]Notable discussions on rationing to which this chapter refers are the following: D. Callahan (1987) *Setting Limits: Medical Goals in an Aging Society,* Simon and Schuster, New York, NY; D. Callahan (1989) Rationing health care: Will it be necessary? *Issues in Law and Medicine* **5,** 3; N. Daniels (1982) Am I my parents' keeper? *Midwest Studies in Philosophy* **7,** 517–540; P. G. Clark (1985) The social allocation of health care resources: Ethical dilemmas in age-group competition. *The Gerontologist* **25,** 2; N. S. Jecker (July 1988) Disenfranchising the elderly from life-extending medical care. *Public Affairs Quarterly* **2,** 3; M. B. Kapp (1989) Rationing health care: Will it be necessary? *Issues in Law and Medicine* **5,** 3; M. A. Somerville (1986) "Should the grandparents die?" Allocation of medical resources with an aging population. *Law, Medicine and Aging* **14,** 3,4; L. R. Churchill (1988) Should we ration health care by age? *Journal of American Geriatrics Society* **36,** 7. Classical reflections by D. Mechanic, C. Fried, H. Hiatt, and J. Childress provide a historical dimension to this topic.

[2]*See* Oregon State Engrossed Senate Bill 27/1989. This action was preceded by a 1987 decision not to fund organ transplantation so that funds could, instead, permit an expansion of the Medicaid program to include more individuals.

[3]L. A. Chung (1990) Rationing health care to the poor. *The San Francisco Chronicle* (September 9). This comment is attributed to M. King, Alameda County (CA) Supervisor, whose county embarked on a similar course.

[4]Chung, quoting A. Caplan, Director, University of Minnesota Center for Biomedical Ethics.

[5]Presidential Commission for the Study of Ethical Problems in Medicine and Biomedical and Behavioral Research (1983) *Securing Access to Health Care,* Washington DC, *see* especially p. 4.

[6]In R. J. Blendon (1989) Three systems: A comparative survey. *Health Management Quarterly* **11 (1),** 2–10, the results of a survey evaluating citizen satisfaction with their country's health care systems are compared. Although there is a feeling that the "system works 'pretty well'" in Canada (56%) and in Great Britain (69%), 89% of US citizens polled believe that "the system requires fundamental change or complete rebuilding."

[7]*See* the result of a major study by the Washington State Medical Association as reported in its 1989 Report of the Executive Committee and Health Access Task Force (Report P, A-89) R. A. Johnson Chair, especially pp. 1–6.

[8]WSMA p. 6.

[9]Draft Criteria for Identifying Priority Health Objectives (December 1989) circulated by the Washington State Board of Health.

[10]However, I concur with R. Priester's assessment that the current preoccupation with cost obscures other issues; *see* his commentary, Health-Care Values Buried by Cost-Control Emphasis *Minnesota Journal* (January 30, 1990) 1,6. I further believe that preoccupation with costs camouflages values that contradict those that our society claims to affirm. The purpose of the next section on Interiority, commencing page 315, illustrates this.

[11]J. McKnight (undated MS) *Do No Harm: A Policy-Maker's Guide to Evaluating Human Services and Their Alternatives.* Center for Urban Affairs, Northwestern University has discussed some of these conflicts tellingly.

[12]I have addressed this topic extensively elsewhere, especially in the following: J. A. Boyajian, ed. (1987) *State of Washington Legal Resource Book* (University of Washington and the Washington Department of Social and Health Services); and J. A. Boyajian, ed. (1989) *Ethical Problems in Aging and Adult Services.* WORKETHICS, Seattle, WA.

[13]*See* R. Priester (1989) *Rethinking Medical Morality: The Ethical Implications of Changes in Health Care Organization, Delivery, and Financing.* Center for Biomedical Ethics/University of Minnesota, Minneapolis.

[14]D. Rowland (1989) Help at home: Long-term care assistance for impaired elderly people. *A Report of The Commonwealth Fund on Elderly People Living Alone.* Baltimore, MD, p. 7.

[15]Rowland, *A Report of The Commonwealth Fund on Elderly People Living Alone.*

[16]Rowland, *A Report of The Commonwealth Fund on Elderly People Living Alone,* pp. 8,37–39.

[17]*See* K. H. Briar and R. Ryan (1986) The anti-institution movement and women caregivers. *Affilia* **1,** 20–31. This important work asks two important questions: Why do women unquestionably accept this role, and how do social workers' assumptions about caregiving and the expectations they place on women express *de facto* sexism? The authors call on their colleagues to examine their personal and collective practice. *See* especially p. 21. I want to remind the reader that the old-old usually are single and, therefore, without immediate family support systems to provide care. Studies indicate that the lack of services to support elders' independent living leads to their physical and emotional neglect, a fact that again affects more women than men. *See,* for example, M. G. Slattery (1988) What lies ahead? *Public Welfare,* **(Spring),** 37,38.

[18]*See* p. 3 in the Employee as caregiver: How business responds. *The Hastings Center Report.* **December 1986.**

[19]Rowland, *A Report of The Commonwealth Fund on Elderly People Living Alone,* p. 8. Furthermore, only Colorado provides a home care allowance that provides a direct payment to a service provider, which is defined as an *informal* or formal caregiver. The member nations of the European Economic Community, however, do provide an allowance to family caregivers.

[20]The social and emotional costs of family caregiving and its relationship to elder abuse and neglect are well documented. For example, *see* N. Hooyman and W. Lustbader (1986) *Taking Care: Supporting Older People and Their Families.* The Free Press, New York, NY; R. S. Wolf and S. Bergman, eds. (1989) *Stress, Conflict and Abuse of the Elderly.* Brookdale Institute of Gerontology and Adult Human Development, Jerusalem; Spring 1988 issue of *Public Welfare* featuring elder abuse.

[21]Briar and Ryan, *Affilia,* p. 21.

[22]Callahan, *Setting Limits,* p. 105.

[23]*See* R. H. Binstock's discussion of "the ageist axioms" that shape policy discussions today in The aged as scapegoat in (1983) *The Gerontologist* **23, 2,** especially pp. 136,137.

[24]Many discount the size of this group, pointing to the small number of Living Wills people have enacted. My own 20 year research with elders in diverse settings support, however, what the polls indicate: this is a large and growing group of people.

[25]S. P. Wallace and C. L. Estes concur, *see* p. 70 (Sept./Oct. 1989) Health policy for the elderly, *Social Science and Modern Society* **26,6:** "Contemporary health policies attempt to 'solve' the health care cost crisis by addressing symptoms through medically based management strategies."

[26]Callahan, Rationing health care, *Issues in Law and Medicine,* p. 354.

[27]*See* L. R. Churchill (1988) Should we ration health care by age? *Journal of the American Geriatrics Society* **36 (7),** 645, which refers to an unpublished work by J. Wax entitled "Ethic and Conflict."

[28]Jecker, Disenfranchising the elderly from life-extending medical care, *Public Affairs Quarterly.*

[29]The latter argument is proposed by Callahan, who argues that when the old give way, they fulfill their role as conservators, for "the primary aspiration" of the old should be "to serve the young and the future..." *See* especially *Setting* p. 43. *See also* Jecker's review of age rationing proposals.

[30]E. Cahn is quoted in J. F. Childress' Who shall live when all cannot live? in R. M. Veatch and R. Branson, eds. (1976) *Ethics and Health Policy,* Ballinger Press, Cambridge, MA, p. 201.

[31]Callahan, Rationing health care, *Issues in Law and Medicine,* pp. 353,354.

[32]S. Sontag (1977) *Illness as Metaphor.* Farrar, Straus and Giroux, New York, NY, p. 82. She further quotes K. Menninger:

> Illness is in part what the world has done to a victim, but in a larger part it is what the victim has done with his [sic] world, and with himself... Such preposterous and dangerous views manage to put the onus of the disease on the patient and not only weaken the patient's ability to understand the range of plausible medical treatment but also, implicitly, direct the patient away from such treatment. (pp. 46,47)

[33]*See* R. P. Preston (1979) *The Dilemmas of Care: The Deformed, the Disabled and the Aged.* Elsevier, New York, NY, pp. 1–7,26.

[34]*See* N. L. Chappell (1988) Society and essentials for well-being: Social policy and the provision of care, in Thornton and Winkler.

[35]As quoted in Kapp, Rationing health care: Will it be necessary? *Issues in Law and Medicine* (Emphasis is added.)

[36]This is not the opportunity for an extensive discussion of academic medicine, which I am developing. For the time being, *see* J. W. Colliton (1989) Academic medicine's changing covenant with society. *Academic Medicine* **64,** 55–60; S. Wolf and B. Berle, eds. (1981) *The Technological Imperative in Medicine.* Plenum, New York, NY; S. A. Schroeder, J. A. Zones, and J. A. Showstack (August 11, 1989) Academic medicine as a public trust. *Journal of the American Medical Association* **262 (6),** 803–811.

[37]Binstock, The aged as scapegoat, *The Gerontologist,* p. 137.

[38]R. Crawford (1977) You are dangerous to your health: The ideology and politics of victim blaming. *International Journal of Health Services* **7 (4),** 664.

[39]J. L. Allen (1979) The inclusive covenant and special covenants. *1979 Selected Papers,* American Society of Christian Ethics, Newton Center, MA, pp. 96–100; *see also* W. F. May's work, especially Code and covenant or philanthropy and contract? in S. J. Reiser, A. J. Dyck, and W. J. Curran, eds. (1977) *Ethics in Medicine.* MIT, Cambridge, MA, pp. 65–76; G. K. Beach (1978) Covenantal ethics, in *The Life of Choice* (C. Kucheman, ed.), Skinner, Boston, pp. 107–125.

[40]Allen, *1979 Selected Papers,* Society of Christian Ethics, p. 113.

[41]*See* J. Moltman's discussion of the vicious circles that disempower in (1974) *The Crucified God.* Harper and Row, New York, NY, pp. 329–332.

[42]M. Farley (1986) *Personal Commitments: Making, Keeping, Breaking.* Harper and Row, New York, NY, p. 72.

[43]Farley, *Personal Commitments: Making, Keeping, Breaking,* p. 107.

[44]I am grateful to B. M. Loomer for using these words in this way. Loomer's principle of size relates to an idea's largeness or smallness—its stature:

> By size I mean...the strength of your spirit to encourage others to become freer in development of their diversity and uniqueness. I mean the power to sustain more complex and enriching tensions. I mean the magnanimity of concern to provide conditions that enable others to increase in stature.

See S-I-Z-E (Spring 1974) *Criterion.* University of Chicago, Chicago, IL, p. 6

[45]Wallace and Estes, Health policy for the elderly, *Social Science and Modern Society,* pp. 66,67.

[46]D. Newquist, (June 1984) Trends in Disability and Health Among Middle-Aged and Older Persons. Andrus Gerontology Center, University of Southern California, CA, p. 66.

[47]*See* B. Jennings (1988) A Grassroots Movement in Bioethics in *A Hastings Center Report Special Supplement* **18:3,** 1–16, for a survey of these efforts.

[48]J. Boyajian (1989) On finding our theological voices: The feminist silence, in *Transforming Thought* (B. B. Hoskins, ed.), Unitarian Universalist Women's Federation, Boston, MA, p. 26.

[49]J. W. Colloton (1989) Academic medicine's changing covenant with society. *Academic Medicine,* **65.**

[50]S. A. Shroeder, J. A. Zones, and J. A. Showstack (1989) Academic medicine as a public trust. *Journal of the American Medical Association* **262(6a),** 803.

[51]A. B. Jonsen (1986) Bentham in a box: Technology assessment and health care allocation. *Law, Medicine and Health Care* **14:3,4,** 172–174.

[52]Kapp, Rationing health care: Will it be necessary? *Issues in Law and Medicine,* p. 351.

[53]E. Bloch (1971) Man as possibility, in *The Future of Hope* (W. H. Capps, ed.), Fortress Press, New York, NY, p. 67.

Philosophical Reflections on Aging and Death

Resentment and the Rights of the Elderly

Albert R. Jonsen

A vigorous debate now rages over the distribution of health care resources in the United States. The place of the elderly in this distribution is hotly contested, not only in the academic discussions, but in the political arena. On the academic side, the problem is framed as a debate over theories of distributive justice; in the political forum, the voice of the voters exerts a power more than theoretical. This chapter suggests a link between the two. A hint of that link appears in a curious passage penned by philosopher David Hume in his *Enquiry Concerning Principles of Morals:*

> Were there a species of creature intermingled with men which, though rational, were possessed of such inferior strength, both of body and mind, that they were incapable of all resistance and could never, upon the highest provocation, make us feel the effects of their resentment: the necessary consequence, I think, is that we should be bound by the laws of humanity to give gentle usage to these creatures, but should not, properly speaking, lie under any restraint of justice with regard to them.[1]

This hypothetical, "were there a species..." is an apt description of the vast population of the frail elderly among us. Although not a species different from ourselves, they are more likely than the general population to be "of inferior strength, both of body and mind" and "incapable of all resistence." Several million elderly dwell

Aging and Ethics Ed.: N. Jecker

in impotence and, perhaps, in impotent resentment, in 6000 nursing homes, requiring assistance in most activities of daily living. Hume asserts that we would not be bound by restraints of justice toward these creatures. This assertion fits his general thinking about justice.

Justice, for Hume, was "not among the natural sentiments of mankind" and "the impressions, which give rise to this sense of justice, are not natural to the mind of man, but arise from artifice and human convention."[2] The artifices and conventions that comprise justice are driven by utility alone: "... And utility aims primarily at the establishment and preservations of the institutions surrounding property." MacIntyre describes Hume's view:

> A system in which pride in houses and other such possessions, and in one's place within a hierarchy, is the keystone of a structure of reciprocity and mutuality, in which property determines rank, and in which law and justice have as their distinctive function the protection of the propertied, so that principles of justice provide no recognizable ground for appeals against the social order...[3]

If justice is conceived as Hume conceives it, the restraints of justice cannot extend to these feeble creatures. Even if they have houses and a place in the social hierarchy, they cannot, because of their "inferior strength of mind and body," take pride in them. They are incapable of participating in a "structure of reciprocity and mutuality." Indeed, utility sees no purpose in surrounding this "species of creature" with the "artifices and conventions" of justice.

However, Hume's paragraph contains a phrase that should stimulate reflection about the place of the elderly in society, apart from his own theory of justice. He writes, "...they could never, upon the highest provocation, make us feel the effects of their resentment." A meditation on the "effects of resentment" might lead us to think somewhat differently than did David Hume about justice and the elderly.

Hume himself counted resentment among the "calm passions," along with benevolence and kindness to children.[4] However, it is not Hume's theory of the passions that inspires this meditation. Rather, the neglected reflections of Max Scheler on the phenomenon of resentment provides the impetus. Scheler, a German philosopher of the first half of this century (1874–1928) wrote his little treatise *Ressentiment* in refutation of Nietzsche's famous thesis, announced in *The Genealogy of Morals,* that impotent hatred and repressed feelings of revenge give rise to the "slave morality" of Christianity. Scheler, although accepting Nietzsche's characterization of *ressentiment* (both philosophers use the French word), denies that it lies at the root of Christian morality. For these two philosophers, resentment is anything but a "calm virtue." Both affirm that this impotent and repressed desire for revenge can generate a moral psychology and, indeed, a morality. Scheler describes the phenomenon:

> Ressentiment is the repeated experiencing and reliving of a particular emotional response against someone else. The continual reliving of the emotion sinks it more deeply into the center of the personality, but concomitantly removes it from the person's zone of action and expression...a suppressed wrath, independent of the ego's activity, which moves obscurely through the mind...In itself it does not contain a specific hostile intention, but nourishes any number of such intentions.[5]

Scheler, in the course of his refutation of Nietzche, provides an exquisite description of the psychology of resentment, its causes, and its effects on personal and social life. In general, resentment falsifies true values and creates false ones, leading to a radical distortion of personal and social life. One who suffers resentment believes and affirms that certain genuine values of which one has been deprived are actually valueless. Scheler wrote, "A person fraught with ressentiment does not admire the earthly...on the contrary, he devalues it. He says, 'all of this is worth nothing, it has no value.'"[6]

Certain states of life and social roles, Scheler notes, are particularly susceptible to the distortions of resentment. Among these is old age. Those who are aging cannot easily resign the values proper to the preceding stage of life and, thus, cannot appreciate the spiritual and intellectual values that remain untouched by the process of aging. They negate and depreciate the values of earlier stages. The old, then, despise the state of being old and equally repudiate the values of youth that they have irretrievably lost.[7]

Scheler says little more about the elderly. We may move on from his text to our own reflections about the resentment of the elderly and its effects on personal and social values. In these reflections, we linger on two images: Hume's picture of a species of impotent and resentful creatures dwelling among us and Scheler's description of the resentment of the elderly distorting personal and social life. The two images do not coincide perfectly, for Hume's picture suggests that resentment is quite passive, churning within the soul but not affecting the world. Scheler makes resentment a powerful, destructive force not only within the soul but in the world: Resentment can be at the root of social structures.

Although Scheler's essay does not study in detail the relationship between resentment and social structure, he alludes to many examples, such as the structure of family, of religious life, and of commercial business. However, one allusion is particularly poignant. He wrote:

> Ressentiment must be strongest in a society like ours, where approximately equal rights (political and otherwise) or formal social equality, publicly recognized, go hand in hand with wide factual differences in power, property and education. While each has the "right" to compare himself with everyone else, he cannot do so in fact. Quite independently of the characters and experiences of individuals, a potent charge of ressentiment is here accumulated by the very structure of society.[8]

The poignancy of that paragraph appears when we recall that "a society like ours" was the society of defeated, inflation-ridden

Germany of the 1920s. The resentment of masses of its population who were promised "equal rights" and "formal social equality," but who found themselves impoverished and powerless, built up the "potent charge" that energized National Socialism and, as a consequence, radically reformed the structures of German life. This is only one example of how the impotent anger of resentment can reshape a society. For Hume, impotence means the absolute inability to exert the power of rank and property; for Scheler, impotence, although equally shorn of that sort of power, is itself a powerful agent of social behavior. It must be reckoned with.

However, is it true to ascribe Hume's picture to the state of the frail elderly in America? Certainly, they are of inferior strength of body and mind; they are incapable of resistance against many assaults, but are they resentful? Also, to what provocations are they subjected? How can we claim that they are provoked to resentment without empirical evidence, polling, and interviews? Might we not find, on actually obtaining such evidence, that the sick elderly are generally content and satisfied or, at least, resigned? Would they, if suddenly endowed with strength of mind and body, revolt against their lot? I cannot make empirical claims about the satisfaction or the dissatisfaction of the elderly.

Even apart from empirical claims, the question might be asked whether the elderly in America are entitled to resentment. Is it a sentiment that they can rightly foster and for which they cannot be blamed? Should we allow that the bitter sentiment of resentment is ever justified? Suggestions of this sort raise the very question that Hume has ruled out, namely, that we might "lie under any restraint of justice" toward the impotent. These suggestions are framed in terms congenial to debates about justice. It is odd, however, to consider resentment an important element in those debates. Yet, on reflection, it may not be so odd.

Resentment is a response to deprivation of some value. More, it is the awareness that the value of which one is deprived is possessed by others. Even more, it is a value that one believes one has a right to. Resentment is, then, a sign of presumed injustice. The deprivation results not merely from chance or from the nature of things, but from social arrangements that block access to the enjoyment of the value. Thus, the elderly cannot resent the loss of physical prowess, although they may regret it. They can resent the loss of prestige or authority that passes from them as they are forced into retirement. The pitiful attempts of superannuated politicians and professors to retain power are vivid evidence of resentment: they actively obstruct good government or good education in their efforts to hang in. Clearly, it would be foolish to claim that the elderly have a right to all they have possessed in their prime, merely because they once had it. Resentment as the loss of these values is certainly not justified. It is not a sign of unfairness.

Still, there may be a deprivation of value that is deeper than the loss of passing youth and its powers, and that constitutes the grounds of justifiable resentment. It is a deprivation to which all who lived a long life are especially vulnerable. All elderly persons have a common characteristic: they have lived through a span of years by repeatedly meeting challenges to survival. Living a life is not passive but active: one stays alive not merely because one's cardiovascular and gastrointestinal systems continue to run on their own, but because one finds ways to nourish the body, to shelter it from the hostile environment, to keep it out of harm's way, and, above all, to find and associate with others who can help one stay alive physically, emotionally, and culturally. Living a life is an achievement. Some persons do it with great vigor and style; others barely make it; yet everyone who survives accomplishes it. The accomplishment deserves acknowledgment.

In many cultures, old age is honorable and honored. This is not an acknowledgment of the personal virtue of individual elderly persons. Even those whose lives may have been clouded with failures or tainted by vice are honored when they reach a certain age. It is the very victory over the multiple challenges of living that deserves honor. It is a commonplace that, in our culture, the elderly are no longer honored. Even Scheler, writing in 1912, noted this:

> In the earliest stages of civilization, old age as such is so highly honored and respected for its experience that ressentiment has hardly any chance to develop. But education spreads...and increasingly replaces the advantage of experience. Younger people displace the old from their positions and professions and push them into the defensive.[9]

The prime justifiable cause of resentment in the elderly is the deprivation of acknowledgment for having lived through a life. Listening to the conversation of the elderly, one frequently hears reference to "how much more difficult it was in my day." Even the "good old days" are praised for the arduous style of life. "Glad we do not have to go through that now, but it made a man out of you." The elderly are inevitably *laudatores temporis acti* (extollers of times past), not because those times were better, but because they were agents who met their challenges and stayed alive through them. It is the ultimate insult to be told that this history was of no account, even more, to be impressed that the younger generation does not care about it, or even know it.

The honor due to those who have lived a life is not manifested in words of praise or in meritorious awards. It consists of being a living part of a society. The Japanese refer to certain elders who have contributed to the arts as "living treasures." Even though no longer actively occupying active roles in social institutions, they are seen as a valued presence in society.

Resentment, then, is the alienation that results from dishonor. It is, as well, a form of injustice. Justice, in its traditional formulation in Western culture, refers to the proper distribution of social goods, among which is honor and dishonor. Modern discussions of justice rarely mention honor (the term is not in the index of Rawl's *Theory of Justice),* yet when Aristotle defines justice, he mentions "distribution of honor" in the first place.[10] There are many reasons, of course, that explain the disinterest in honor as a subject of justice. Without expanding on them, we may say that the problem of justice to the elderly is rendered doubly difficult because honor is omitted. Seeing justice as almost exclusively a matter of distribution of material goods prejudices the discussion. It accepts without question Hume's premise that justice is a matter of utility in the distribution and maintenance of property.

Can we entertain the Aristotelian notion that distribution of honor is prior to distribution of any other goods in social life? In so doing, we do rejoin certain modern streams of thought that see respect for all persons and the support of a sense of personal worth and dignity as radical, primary goods (a consideration that is found in Rawls' pages). The arrangements of social institutions that distribute powers, entitlements, and material goods are to be structured around this more radical human good. For the elderly, the quality of personal worth, and the object of honor are the achievement of having lived a life. It is common to all who have survived and is worthy of honor.

However, renewing interest in honor may make the discussion more, rather than less, complex. Honor is associated with desert. Desert is another subject that has been given slight attention in modern theories of justice. One of the few scholars to reflect on honor and desert, Michael Walzer, remarks, apropos of Rawls, "advocates of equality have often felt compelled to deny the reality of desert."[11] However, the honor of the elderly is something of a paradox: it is an honor of which all are equally deserving when old. Just as we have, since the Revolutionary era, depreciated the honor pertaining to royal or noble status, so now we depreciate any honors as attaching to a class as such. Honors are to be rendered, if at all, by strict and honest evaluation of differential merit.

However, can it not be seriously entertained that honor and desert need not be differential, at least in some aspect of life? Does this not accord with certain social conventions: we say, "I respect the clergy (or doctors) even though some of them are disreputable." Is it not reasonable to say, as many past generations and other cultures do say, "One must respect one's elders," without differentiating on grounds of greater or lesser achievement?

It is, perhaps, because the work of living a life is so intimately personal and so uniquely individual that we can dispense with differential evaluation. No one but the one who has lived the life knows the challenges and the efforts required to meet them. Few persons have a biography written about them, but all persons write their own biographies in the inner words of challenge and response. Even the outward events that can be judged as successes or failures by observers are not susceptible to evaluation in motive, effort, or internal obstacle. Thus, the honor of having achieved a life can be distributed equally uniformly.

Just resentment arises, then, from uniform deprivation of honor to the elderly. What has this to do with the social arrangements that distribute income, housing, health care, food, household help, and the other goods that figure more prominently than honor in discussions of justice for the elderly? Certainly, none of the problems of differential distribution of these social goods across generations are solved by affirming the honorable place of the elderly. Honoring one's elders does not necessarily entail providing them all with coronary artery bypasses and hemodialysis *in perpetuum.* It does not dictate the priorities between funding elder day care over infant day care. If so, may not "Old Jack Falstaff" have spoken most truly: "what is honor then? A word. What is that word, honor? Air. Who hath it? He that died o' Wednesday. Doth he feel it? No. Doth he hear it? No. It is insensible, then? Yea, to the dead."[12] Honor is cold comfort to those who are left out of the distributions of social goods on grounds of just and fair allocations. Like all basic principles, it may be too basic to be of benefit when other competing claims are acknowledged.

If the institutions of a society can be so arranged as to mitigate resentment, whatever else they do, justice will be done to the elderly. Honor becomes sensible, rather than, as Falstaff said, "insensible" when it comes to the living in the form of social recognition of their need to live, while they live, as independently and as decently as possible. The persistent complaint that elderly persons are often forced to liquidate all assets, including their homes, in order to pay the exorbitant costs of long-term custodial care illustrates the "insensibility" of public policy. Anyone who has lived through a life successfully enough to possess a home and some means is justly resentful of being beggared by the costs of caring for a spouse. Similarly, persons who are forced to sell homes they have lived in for many years because they are unable to pay notably increased property taxes can be justly resentful. Again, policies that reduce financial, retirement, or health benefits for which elderly persons have contributed, and have been led to expect, can stimulate just resentment. These are deprivations of goods that have been won by overcoming the challenges of life.

In light of these considerations, it seems strange to concede Hume's denial of justice to those who are impotent to enforce their resentment. The resentment that is stimulated from loss or lack of due honor is justly felt. The social institutions and practices that deprive persons of due honor are unjust. Similarly, tolerance of these social institutions and practices is tolerance of injustice. Due honor is justly distributed to those who have met the challenges of living a life: it consists of creating and fostering the social institutions that make life possible without resentment.

The first paragraph of this reflection noted a link between the theoretical and the political problem of justice. That link is found in the way in which resentment manifests itself in a democratic society. Impotent though the frail elderly may be, they exercise an influence over others who are not quite as impotent, the elderly voter. That influence lies in the vision of becoming dishonored merely through the passage of a few years of time. In a democracy, especially one in which an increasing number of voters will be in the upper age brackets, those who foresee dishonor and begin to feel the resentment associated with it are not impotent. They are able to exercise the power of the vote and may do so in ways that manifest resentment more than political prudence. The recent repeal of the catastrophic health insurance provisions may illustrate this: the powerful voice of the lobby of the elderly was swayed by the injustice of the elderly alone being taxed to finance their care, but, in so doing, their resentment forgot those elderly who have been benefited without being taxed. Resentment must be reckoned with, lest it distort values and make even worse the condition of those who suffer its ravages.

This reflection ends, not with the Scot Hume nor with the German Scheler, but with the Frenchman Montaigne. His wise remarks on the vicissitudes of old age balance a realistic perception of the phenomenon of resentment with his personal optimism. He wrote, "I find in old age an increase of envy, injustice and malice. It stamps more wrinkles on our minds than on our faces..."[13] The wrinkles on the mind distort the view of the world. However, the wrinkles may come, not so much from old age as such, as Montaigne suggests, but from the rough treatment the world has given to persons who have met its challenges over many years. At least to the extent that these wrinkles on the mind are the effects of dishonor and of righteous resentment, it is a matter of justice to smooth them out by just and fair social arrangements.

Notes and References

[1]D. Hume (1965) Enquiry in *Hume's Ethical Writings* (A. MacIntyre, ed.), Collier-Macmillan, London, UK, III, i. p. 41.

[2]Hume, Enquiry, *Hume's Ethical Writings.*

[3]A. MacIntyre (1988) *Whose Justice? Which Rationality?* University of Notre Dame Press, Notre Dame, IN, p. 295.

[4]Hume (1888) *Treatise of Human Nature* (L. A. Selby-Bigge, ed.), The Clarendon Press, Oxford, UK, II, iii, p. 423.

[5]M. Scheler (1961) *Ressentiment* (L. A. Coser, ed.), The Free Press, New York, NY, pp. 39,40.

[6]Scheler (1973) *Formalism in Ethics and the Non-Formal Ethic of Values*, Northwestern University Press, Evanston, IL, p. 231.

[7]Scheler (1961), pp. 62,63.

[8]Scheler (1961), p. 50.

[9]Scheler (1961), p. 63.

[10]Aristotle (1962) *Nicomachean Ethics,* V, ii, 1130[b], (M. Ostwald, ed.), Bobbs-Merrill Company, Indianapolis, IN, p. 117.

[11]M. Walzer (1983) *Spheres of Justice*, Basic Books, New York, NY, p. 260.

[12]Shakespeare, *Henry IV*, Part I, V, i.

[13]M. Montaigne (1958) *Essays* (J. M. Cohen, trans.), Penguin Books, London, UK, III, ii, pp. 249,250.

Ancient Myth and Modern Medicine

Lessons from Baucis and Philemon

Lawrence J. Schneiderman

The Myth

Here is the story of Baucis and Philemon as recorded in *Bullfinch's Mythology:*[1]

> On a certain hill in Phrygia stands a linden tree and an oak, enclosed by a low wall. Not far from the spot is a marsh, formerly good habitable land, but now indented with pools, the resort of fen-birds and cormorants. Once on a time Jupiter, in human shape, visited this country, and with him his son Mercury (he of the caduceus), without his wings. They presented themselves, as weary travelers, at many a door, seeking rest and shelter, but found all closed, for it was late, and the inhospitable inhabitants would not rouse themselves to open for their reception. At last a humble mansion received them, a small thatched cottage, where Baucis, a pious old dame, and her husband Philemon, united when young, had grown old together. Not ashamed of their poverty, they made it endurable by moderate desires and kind dispositions. One need not look there for master or for servant; they two were the whole household, master and servant alike. When the two heavenly guests crossed the humble threshold, and bowed their heads to

pass under the low door, the old man placed a seat, on which Baucis, bustling and attentive, spread a cloth, and begged them to sit down. Then she raked out the coals from the ashes, and kindled up a fire, fed it with leaves and dry bark, and with her scanty breath blew it into a flame. She brought out of a corner split sticks and dry branches, broke them up, and placed them under the small kettle. Her husband collected some pot-herbs in the garden, and she shred them from the stalks, and prepared them the pot. He reached down with a forked stick a flitch of bacon hanging in the chimney, cut a small piece, and put it in the pot to boil with the herbs, setting away the rest for another time. A beechen bowl was filled with warm water, that their guests might wash. While all was doing they beguiled the time with conversation.

On the bench designed for the guests was laid a cushion stuffed with sea-weed; and a cloth, only produced on great occasions, but ancient and coarse enough, was spread over that. The old lady, with her apron on, with trembling hands set the table. One leg was shorter than the rest, but a piece of slate put under restored the level. When fixed, she rubbed the table down with some sweet-smelling herbs. Upon it she set some of chaste Minerva's olives, some cornel berries preserved in vinegar, and added radishes and cheese, with eggs lightly cooked in the ashes. All were served in earthen dishes, and an earthenware pitcher, with wooden cups, stood beside them. When all was ready, the stew, smoking hot, was set on the table. Some wine, not of the oldest, was added; and for dessert, apples and wild honey; and over and above all, friendly faces, and simple but hearty welcome.

Now while the repast proceeded, the old folks were astonished to see that the wine, as fast as it was poured out, renewed itself in the pitcher, of its own accord. Struck with terror, Baucis and Philemon recognized their heavenly guests, fell on their knees, and with clasped hands implored forgiveness for their poor entertainment. There was an old goose, which they kept as the guardian of their humble cottage; and they bethought them to make this a sacrifice in honor of their guests. But the goose, too nimble, with the aid of feet and wings, for the old folks, eluded their pursuit, and at last took shelter between the gods themselves. They forbade it to be slain; and spoke in these words: "We are gods. This inhospitable village shall pay the penalty of its impiety; you alone shall go free from the chastisement. Quit your house, and come with us to the top of yonder hill." They hastened to obey, and, staff in hand, laboured up the steep ascent. They had reached to within an arrow's flight of the top, when, turning their eyes below, they beheld all the country sunk in a lake, only their own house left standing. While they gazed with wonder at the sight, and lamented the fate of their neighbors, that old house of theirs was changed into a temple. Columns took the place of the corner posts, the thatch grew yellow and appeared a gilded roof, the floors became marble, the doors were enriched with carving and ornaments of gold. Then spoke Jupiter in benignant accents: "Excellent old man, and woman worthy of such a husband, speak, tell us our wishes; what favour have you to ask of us?" Philemon took counsel with Baucis a few moments; then declared to the gods their united wish. "We ask to be priests and guardians of this your temple; and since here we have passed our lives in love and concord, we wish that one and the same hour may take us both from life, that I may not live to see her grave, nor be laid in my own by her." Their prayer was granted. They were the keepers of the temple as long as they lived. When grown very old, as they stood one day before the steps of the sacred edifice, and were telling the story of the place, Baucis saw Philemon begin to put forth leaves, and old Philemon saw Baucis changing in like manner. And now a leafy crown had grown over their heads, while exchanging parting words, as long as they could speak. "Farewell, dear spouse," they said, together, and at the same moment the bark closed over their mouths. The Tyanean shepherd still shows the two trees, standing side by side, made out of the two good old people.

Exploring the Myth Today

Is the examination of this ancient myth by a modern physician anything more than a high-minded literary diversion, an obligatory nod to the Humanities? At first glance it might seem so. The modern physician knows better than to take anecdotal evidence seriously. Indeed many of the miracles of modern medicine have been achieved by overthrowing the myths of the past, replacing them with dogged empiricism. Admittedly, medicine *is* infiltrated with moral questions, and true, the ancient Greeks *did* have some interesting moral answers back when life was simple, but what have they done for us lately? What can they teach us about organ transplantation, gene therapy, artificial insemination, DNR orders, prolongation of vegetative existence, or allocation of intensive care unit beds? Not only medicine, but *life* has become so much more variegated and complex. Unlike ancient Athenians, we live in a vast country, derived from a plethora of cultures, sharing no common religion, and constantly interacting with an even vaster, more complex world. Take, for example, that great classical art form, drama. Unlike today's audience, which demands novelty, astonishment, and *variety,* from the playwright, the ancient Greeks flocked to the ceremonies of drama to experience the familiar. A single plot served many plays, and a single myth any number of playwrights, including the celebrated ones—Sophocles, Euripides, Aeschylus, and Aristophanes. What would compare in these more pluralistic and secular times? George Washington and the Cherry Tree dramatized by Arthur Miller, David Mamet, Sam Shepard, Marsha Norman, and Neil Simon? Obviously, the role of such legends in society—except as sources for fluffy musicals—has greatly diminished. Have they entirely lost their relevance?

Loneliness, Unequal Passage Through Life, Vulnerability to Fate

I remember when the myth of Baucis and Philemon first entered my medical ruminations. I was seeing Mr. and Mrs. D in my office. Both were in their 70s. Fate had treated him far more kindly than her, however. She was definitely in the category of "looking older than her stated age"—frail, requiring help for her simplest everyday tasks, and utterly terrified at the thought of leaving the security of her home. Mr. D, on the other hand, was still physically vigorous, bright, cheerful, and extroverted. He had conceived the fanciful notion of traveling continuously around the world in whatever years were left to him, but Mrs. D was not up to it. I could see their marriage vows of "for better or for worse" and "in sickness and in health" were being sorely tested. Although applying the most devoted attentions to his wife, Mr. D was chafing at the restrictions on his life, and for Mrs. D, this only added guilt to her suffering. Early in my practice I had come upon a terrible secret—the pain lurking for two people bound by love, made all the worse by love, in their unequal passage through life. That was when I remembered the wish that Baucis and Philemon had made to the gods: "...[S]ince here we have passed our lives in love and concord, we wish that one and the same hour take us both from life, that I may not live to see her grave, nor be laid in my own by her." What struck me was that the two mythological characters had asked, not for health, beauty, riches, long life, or any other miraculous gift the gods presumably were up to providing, but merely to be granted a simultaneous death. Since then, I have discovered that this terrible secret—the desirability of timing one person's death to match anothers—is no secret to the elderly.

Mr. and Mrs. B, now in their 80s, have been discussing their future with me for all the nearly 20 years they have been in my care. At first, he was the one whose health was more precarious. When they came to me, he already had diabetes, high blood pressure, painful arthritis, a hernia, and symptomatic prostate trouble. She, on the other hand, was sharp-witted, socially active, energetic, and intent that the two of them continue to "use it and not lose it"—definitely the more promising of the two. We frankly addressed their prospects under the reasonable assumption that Mrs. B would be the caretaker and ultimate survivor, a calculation made all the more likely after Mr. B developed atrial fibrillation, heart failure, and bilateral atherosclerotic obstruction of his femoral arteries, requiring an aortic bypass procedure. Would she be able to take care of him, I wondered, and how would she get along when he was gone? Astonishingly, he came through all his surgery, experienced a revival in his spirits and physical ability, and even took up an exercise program of golf several times a week. In the meantime, Mrs. B began to show the earliest signs of a strange, relentless neurological deterioration resembling Parkinson's disease—it was she whose life suddenly took a downward course. Soon her husband had to assist her with walking, scurrying about as best he could to protect this willful—to the point of reckless—woman from falling. When we managed to persuade her to use a wheelchair, it allowed him to push her around the retirement village; even though she could barely speak, at least she could listen to conversations he engaged in with other residents. However, now in this last year, *his* career has taken a turn for the worse. *He* was the one who fell and fractured his spine, and despite medications, his various illnesses, including progressive heart failure, have now forced him to abandon his golf game and limit his walking to short distances. Yet despite all his own problems, Mr. B continues to ask with the greatest feeling, "How am I going to take care of her?" It is clear that their long lives together are coming to an end, but who could have predicted this spiraling trajectory? Both agree—and openly discuss—if she dies first, it will not be as bad as if he dies first. Where would she go? What would happen to her? Their children could not give her the total attention Mr. B gives her. She would rather die than go to a nursing home. Indeed, every utterance seems to echo Baucis and Philemon: "We wish that one and the same hour may take us both from life." What would the American Association of Retired Persons say if they could hear her speak? They would clearly be alarmed. She meets the suicide "warning signs" of the AARP's Public Policy Institute: social isolation and loneliness; a history of recent losses and intractable pain; change in status related to income, employment, and independence; and fear of institutionalization in a nursing home. Yet what does the myth say? It says that a good and timely death no less than a good and timely life is a gift of the gods.

Is it proof of society's advancement that our modern gods would not grant Baucis and Philemon their wish?

Yielding, Passing on

The term "passing on," a much mocked euphemism for dying, can also represent the obvious fact that, when we die, we pass on what we have possessed to those who follow us. Baucis and Philemon did not ask to join the immortal gods, but rather "to be priests and guardians of this your temple"—in other words, to remain on earth and be the temporary stewards of the symbols of their benedictions. The elderly are no different from anyone else in wanting to enjoy their lives to the fullest, but they do dread outstaying their welcome, becoming a burden, and leaving nothing but bitterness as their memorial. Not only from personal economic considerations do they object to exhausting their life savings on useless medical treatment, but also from the desire to pass on to children and grandchildren property, acquisitions, memorabilia, and monetary and other legacies—*proofs* of their existence. It is a common experience that, at some point, a dying person accepts the inevitable. Unlike the young (and animals) who never imagine themselves growing old, much less dying, the elderly contemplate their deaths in considerable detail—as is evident by any survey that has been done regarding wishes for terminal care. My own clinical experience, as well as clinical research, indicates that the majority of elderly patients do not want aggressive medical measures used to keep them alive if they become permanently unconscious, for example, and do not want life-support treatments if that means living permanently attached to a ventilator, or requiring artificial nutrition and hydration in the hospital. At some point, they *want* to pass on.

Reduction to Essentials, Value of Simplicity

How exiguous is the world of Baucis and Philemon. The threshold the two heavenly guests cross is "humble." They eat out of "earthen" dishes and "wooden" cups on a cloth that is "ancient and coarse." The goose they attempt to catch and serve is old and their last. Even the rickety table needs the prosthetic assistance of a piece of slate. Every detail serves to underscore that which today would be called their marginal socioeconomic state. Yet "not ashamed of their poverty, they made it durable by moderate fires and kind dispositions."

Yesterday, most of my time with Mrs. R was spent trying to cheer her up and help her figure out ways to pass the time. She admits to being old (she is in her late 70s), but not to being ill (except for dizzy spells that prevent her from driving her car), only miserably bored, which she would not be—indeed she would have no complaints at all—if only she could just *go* places—local museums, the library, art galleries, a few concerts and lectures, and a few senior citizen and Sierra club activities. These simple capacities would be all it would take to improve those measures of health social researchers love to collect—activity level, quality of well-being, locus of control, sense of coherence, and so on. However, this city, in which two hospitals are jostling to promote their heart-lung transplant programs, provides her no affordable way to get around. The public transportation and Dial-a-Ride systems are pathetically inadequate. Our governmental gods could grant her wish if it were for a heart-lung transplant. Medicare would pay the hundreds of thousands of dollars for the procedure, plus followup drugs, biopsies, specialists, and complex followup care. However, the gods have ordained no medical insurance that pays for taxicabs. Thus in an age of technological splendor, Mrs. R suffers from the simplest forms of deprivation—which as her physician I try to help make durable by "moderate fires" (of conversation) and "kind [and frustrated] dispositions."

Images of Nature's Cycles and Seasons

One day I noticed tears in the eyes of Mrs. B, the elderly patient suffering from the Parkinson's-like disease. Her husband was describing their activities since I had last seen them. I interrupted to ask how she felt. Haltingly, but with the force of despair she said, "I feel life has betrayed me."

I could picture her life somewhat since I had once made a housecall to get an idea of where she lived, and in particular to help the two of them identify and minimize injury hazards. As I wandered around her retirement complex, I could not help thinking again of Baucis and Philemon—exchanging parting words, putting forth leaves, metamorphosing into a linden tree and an oak—climaxing a story in which images of transforming nature abound. From the very beginning, the story tells of time's passage, and how it causes not only decay, but replenishment. The setting is "a marsh, normally good habitable land, but now indented with pools, the resort of fen-birds and cormorants." Wine is miraculously restored in the pitcher, a flood engulfs the village, and their house is transformed into an opulent temple, echoing other replenishment myths—Christ and the wine, Noah and the Ark, to name just two. Life ends, and life renews, the story says.

In contrast, there are no reassuring images of nature's renewal to comfort Mrs. B. The retirement village is trimly landscaped with low-maintenance shrubs bred to the perfection of artificial flowers. Rarely do they drop a leaf onto the tidy banks of redwood chips. The materials of her world consist of glabrous stucco, tile, wrought iron, and molded plastic furniture. The chlorinated pool is not the fertile resort of fen-birds and cormorants, but only of the regulars who do their daily laps. Instead of a muddy path scattered with surprises, she has glaring swathes of concrete; instead of rocky hillsides, she has the comforts of elevators; instead of weather she has air conditioning. The slightest hint of decay is hastily repaired, and nothing intrudes into this well maintained blandness. Life for her is an unchanging schedule of card games she cannot join, conversations she can only listen to, and exercise classes she can only watch. As senior citizens residences go, it is considered one of the best.

Society's Care of the Aged, Their "Entitlement" Required by the Gods

We cannot know for certain, but the gods' appointment of Baucis and Philemon as "keeper of the temple as long as they live" suggests that the elderly played a vital, even powerful role close to the central symbols of religious worship in classical Greek society. From our utilitarian perspective, we might wonder why they were given such a place of honor in a period of history that surely must have been dominated by want, hardship, and uncertainty, not to mention the requirements of hard physical labor, propagation, and defense. We conjecture, of course, that the elderly *were* a vital link to the past. Before printing presses and computers, the only way to store lessons of survival was in the recollections of those who had experienced crises, threats from enemies, and natural disasters—and survived. Archives and libraries reposed in people's memories. The same would likely be true of myths and religious rituals, which could be transmitted to novitiates only by those versed in their complexity.

Today the elderly play no such role in society and, consequently, enjoy no such honors. Rather, they are objects of patronizing debate regarding the proper allocation of responsibility and resources for their care. The best hope the elderly have is that the debate will be "compassionate"—that as hapless victims and refugees in the war between the elderly and the young and between the haves and the have-nots, they will be treated humanely.

For today, to be both elderly and poor is to have two reasons for being socially extraneous and burdensome. The elderly, says the philosopher Daniel Callahan in a book entitled *Setting Limits*[2]—a term parents use on pets and fractious toddlers—are not "socially *indispensable* in the way that children and young adults are for a society" (italics his). Although the book is a deeply sympathetic meditation on the role of the elderly in society, by entering the lists in the midst of the current debate on health care cost control, it inescapably confirms the underlying assumption of its title: namely that the elderly are being indulged. It is a view that most people are unwilling to acknowledge and Callahan has been roundly criticized for impolitically exposing. Yet, ironically the fact that we as a nation contribute so generously to the elderly's welfare may actually be a measure of our compassion. For, as opposed to ancient societies, we have so little use for them. Merely look around and you can see that youth rules today: loud music; flamboyant fashions; glossy magazines; brash, unreflective, iconoclastic movies and television; and general impatience. Any hope that the musings of the elderly might be heard, much less taken seriously, even much less considered vital to society's survival or religious significance suggests delusions of grandeur. Yet one cannot help wondering what they would prefer: to be supported on grounds of compassion or—as were Baucis and Philemon—on the grounds of social necessity.

Communal Ethics

Finally the story asserts a communal ethic. Jupiter and Mercury presenting as weary travelers were rejected by all the village inhabitants until they were welcomed by Baucis and Philemon. In ancient societies, which lacked organized charities, welfare systems, and various forms of disability insurance (health insurance and unemployment insurance), the individual relied (like Tennessee Williams' Blanche DuBois) "on the kindness of strangers." It should be no puzzle that inhospitality was harshly punished. A lonely traveler had only those limited means for survival as the landscape afforded. Life could depend on it. In a time of precarious nature and capricious fortune, each person was utterly beholden to the good will and charity of others. This is something the healthy and vigorous learn when they get sick, and the young learn when they get old. Like any other doctor caring for elderly patients, I realize that nothing I do in my office compares to what others in the community do outside. Without followup care, all the most miraculous medicine and surgery will fail. How sad that the elderly must suffer until society makes this discovery—that it is the devotion of the *community* the elderly require more than the virtuosity of any one individual.

The debate over recent efforts to supplement Medicare with "catastrophic care" coverage illustrates this problem. Only an estimated 10% of the elderly were predicted to require prolonged medical treatment and hospitalization. Indeed, that could well be an overestimate, since (as I pointed out earlier) it is likely that, if offered the opportunity to make better use of advance directives, such as Living Wills and Durable Powers of Attorney, substantially more of the elderly would reject high-technology aggressive treatment than receive it now. Although the exact reasons for the outcry and protests against the plan by the elderly are not clear—and sadly do seem to reflect the war between the haves and the have-nots and the war between the elderly (who were being exclusively taxed) and the young (who were allowed to escape sharing the burden)—one major source of objection is that what the elderly really want is coverage for long-term chronic—including nursing home—care. It is a further irony that the administration's objection to repealing the catastrophic care plan was not because it failed to deliver the kind of medical care the elderly wanted, or the kind of care the elderly needed, or anything else that had to do with the health of the elderly, but rather because the surtax had already been calculated as a short-term revenue source in the battle to reduce the budget deficit without "raising taxes." (Apparently raising the surcharge on Medicare escaped the administration's definition of taxes.) Thus, the health care of the elderly was being debated not on its merits, but as another ploy in deceptive budgetary manipulation.

What would Jupiter and Mercury have made of our government's inhospitality? Would they have ordained that it "pay the penalty of its impiety"? If our country suffers a similar fate of that wracked village before the eyes of some wondering Baucis and

Philemon—perhaps an ever-deeper immersion in divisive social and economic bitterness—it could be because we failed to heed the lessons of this ancient myth.

References

[1]*Bullfinch's Mythology* (1939) Carlton House, New York, NY, pp. 44–46.
[2]D. Callahan (1987) *Setting Limits.* Simon and Schuster Inc., New York, NY, p. 36.

The Meaning of Temporality in Old Age

Nancy S. Jecker

The meaning of aging for the individual is shaped from and legitimated by a distinct time concept. The time concept through which individuals interpret their temporal world itself relates to the broader conceptual repertoire about time we entertain as a society. This chapter first makes explicit alternative models of time, and relates these to notions of subjective and chronological aging. I explore the values implicit in these models, and consider alternative approaches to the significance of time and aging.

Time Concepts

"Time" comes from the Sanskrit root meaning "to light" or "to burn." It is also the basis of the Latin "tempus," the Italian "tempo," and the English "temperature" and "temperament," all words having vibrational frequencies and suggesting that time is the fire of life.[1] In ancient Greece, the word for time also represented the word for life.[2] In Western philosophy, two dominant models of time support different assessments of the meaning of time and aging. Dating back to Aristotle, one view identifies time as an objective and quantifiable entity, ontologically prior to our consciousness of it. Aristotle related time to motion, arguing that "time is a measure of motion and of being moved."[3] Likewise, his definition of motion presupposed temporal measure: "not only do we measure the movement by the time, but also the time by the movement, because they define each other."[4] According to this conception, "there is the same time everywhere at once,"[5] and time remains unaffected by our choice of measuring instrument. Carrying forward the Aristotelian tradition in modern times, Newton described time as arising from physical processes independent of a perceiving subject. Newton held that the laws of motion presuppose the existence of an absolute space and time in which bodies can truly be said to be in motion.[6] According to him, absolute time flows without regard to external events and is distinguished from the measures of time we ordinarily make that are relative to material reference frames.

As aging marks the passage of time, this first picture of time forms the basis for experiencing aging in a distinct way. On this model, the individual's movement through time appears to be externally imposed, rather than emanating from within the person; unstoppable, as opposed to regulated by biological events or life choices; and objectively real, rather than subjectively defined. Reflecting this tradition, much of the early gerontological literature addressing time consciousness in aging persons focused on subjects' estimations of time intervals compared to clock or calender time, and defined individuals' temporal perspectives as past or future oriented relative to a fixed standard. Commenting on this approach, Hendricks and Hendricks[7] note that gerontologists traditionally have taken for granted a quantitative and linear approach to time, an approach that views time as external to the interview subject. Time is "seen as being analogous to points on a line, in both direction and order; it is a unique standard by which everyone can be compared."[8]

An alternative time concept has roots in St. Augustine and in the Middle Ages. Augustine construed time to be both qualitative and paradoxical. For him, time is qualitative, rather than quantitative, because it depends on the existence of God and the material world, and because its motion is not defined in terms of any absolute measure. "Do you command me to agree with someone who says that time is the movement of a body?" asks Augustine, and he replies, "You do not command this."[9] Time itself is immeasurable; we measure only our own actions in time through images of times past, present, and future that reside in our mind or soul. According to Augustine, "It is in you, O my mind, that I measure my times."[10] The paradoxical nature of time is revealed by the fact that, in any attempt to measure time, part of the process to be measured has passed away, and part of it has yet to be. In other words, it is impossible to place the thing to be measured before me all at once, as is possible with other measurements, such as length.[11] In contemporary metaphysical and scientific writings, a perspective similar to Augustine's is found. Rather than regarding time as a real thing, within which the whole succession of natural events in the world has a definite position, contemporary philosophers generally treat time as a relational concept. Thus, the duration and temporal location of events are relative to other events and persons. This view became a dominant philosophical outlook following Einstein's discovery that concepts of absolute time must be modified for rapidly moving bodies. Eventually, this idea was extended to include frames of reference in all types of relative motion.[12]

The experience of aging that grows out of this second temporal model marks time by relative cues, such as the age norms of others and of the larger society. Social norms for aging persons generally serve as the reference that cues individual older persons about how to interpret their own experience of time and aging. Reflecting this approach, gerontologists increasingly have focused attention on subjective aging, defined as how young or old elderly individuals *perceive* themselves to be. For example, an individual's subjective age may be set by stating, "We would like to know how old you feel," and then asking, "Would you say you feel young, middle aged, old, or very old?"[13] Subjective age has been compared with "real" or chronological age, and a greater discrepancy than would be expected has been found between individuals' perceptions of chronological age and their actual chronological ages. Yet during old age, individuals also possess a highly refined concept of the "normal, expectable life cycle," defined as "a set of anticipations that certain life events will occur at certain times, and a mental clock telling them where they are and whether they are on time or off time."[14]

Time and Meaning

Because these contrasting models of time shape our personal experience of aging, it is important to lay bare the values and norms they impart to personal aging. Despite their noted differences, both models share important value assumptions. On both accounts, the tendency is to reify time and so wrest from individuals the ability to construct unique temporal worlds. This is most clear on the Aristotelian model, where time is understood as an extant, quantifiable entity. Clock time stands as an absolute marker of temporal flow—a yardstick by which the individual's temporal experience is measured. Yet the Augustinian model, which depicts time as relative to external cues, also posits a "normal human experience" of time. Augustine's reference to images of time that reside in the mind or soul presupposes that different minds and souls share common images, and the Einsteinian picture of time also takes for granted that individuals possess a common temporal reference frame. On both models, unless a qualifying condition is present, similar temporal experiences are assumed and departures from this are affixed a label of abnormality.

A second feature found on both accounts is that certain sensations, such as excessive daydreaming, dwelling in past time, perception of the temporal flow as discontinuous, and memories of greater impact than our present perceptions, constitute abnormal temporal perspectives.[15] Remarking on the significance of these assumptions in gerontological research, Kastenbaum speculates that

> ...perhaps what irritates us...about an elderly person's past orientation is our own reaction of having been snubbed. The present moment is consensual, public; we all share in it...so the person who dwells on the present or the future has company. But when

> an aged person dwells on his past he is moving in a realm that is not directly accessible to us; we were not a part of it.[16]

Whether from a clock or a social consensus, external models of time create public norms that shape and limit individuals' interpretations of aging and temporal experiences.

Such standards have the potential to become particularly onerous for elderly persons, whose temporal pace and orientation may depart from norms present in the population at large. For example, researchers report that a distinct bias toward specifying a younger subjective age exists among older age groups.[17] Whereas feeling older and developmentally "ahead" may prompt positive self-assessments in early years, researchers have noted that in old age reporting a *younger* subjective age is associated with positive variables, such as general life satisfaction,[18,19] improved psychological functioning,[20] and maintaining a meaningful sense of purpose and existence.[21] In addition, a healthy response to aging may include a sense of "oneness with the universe and the flow of time"[22] or a sense of timelessness.[23] Finally, in old age, coping with mortality may call on skill at interweaving past, present, and future into an integrated whole. If so, then allowing the elderly person latitude in thinking about eternity or experiencing time's flow may enhance self-identity and integration of life events.[24]

These remarks suggest that in old age, and perhaps in other circumstances as well, focusing on the relentless movement of time and distinguishing clearly between past, present, and future may impede self-identity and enjoyment. One explanation for this may be that such a temporal rhythm serves to banish questions about the meaning of life or the place of old age in a meaningful life. If so, then those who strive to retain the temporal perspective held in youth or middle age may experience old age as being simply one more stage, rather than a unique stage with distinct and daunting tasks. By contrast, those who conceive of all ages as part of a larger whole and who cultivate their own temporal rhythms may come to see the whole of life with broader vision. They may interpret what came before with reference to this whole. Old age lived in this way may even glimpse beyond one's own existence, and view the future time in which one will not be with a sense of greater ease and closure.

Broader Implications

The alternative time consciousness I am suggesting should aim to be functional for diverse groups and individuals. It should measure time by internal cues, such as personal biography and individual biology, rather than external or social time standards. Finally, it should leave open the possibility of "time travel" or subjective movement backwards and forwards through time. In this way, ample room is made for older persons' variances in temporal experiences.

The broader implications of this approach are striking, for it is out of our most fundamental concepts of time and timing that models of health and disease, and theories of psychological development are built. As individuals, we are influenced heavily by these models and measure our own success in terms of progress along popularized versions of life-cycle theories. Our self-esteem and confidence are supported or squelched by the results.

Supporting unique standards of time and timing for aging persons need not imply "losing touch" with real time, nor need it entail displacing the philosophical models of time we have in place. Instead, it calls for acknowledging the values implicit in these models and making reasoned choices about their validity for particular aging persons. By supporting unique temporal experiences, we allow the play of past memories, the anticipation of future events, and the living through present to become a more authentic and adaptive experience for aging persons.

References

[1]N. E. Strumpf (1986) Studying the language of time. *Journal of Gerontological Nursing* **12,** 22–26.

[2]Strumpf, Studying the language of time, *Journal of Gerontological Nursing,* pp. 22–26.

[3]Aristotle (1941) *Physics,* in *The Basic Works of Aristotle* (R. McKeon, ed.), Random House, New York, NY, pp. 218–397, at 294.

[4]Aristotle, *Physics,* in *The Basic Works of Aristotle,* p. 294.

[5]*Physics,* in *The Basic Works of Aristotle,* p. 294.

[6]D. Shapere (1967) Newtonian mechanics and mechanical explanation, in *The Encyclopedia of Philosophy* (P. Edwards, ed.), McMillan Publishing Company, New York, NY, pp. 491–496.

[7]C. K. Hendricks and J. Hendricks (1976) Concepts of time and temporal construction among the aged, with implications for research, in *Time, Roles, and Self in Old Age* (J. F. Gubrium, ed.), Human Sciences Press, New York, NY, pp. 13–49.

[8]Hendricks and Hendricks, Concepts of time and temporal construction among the aged, with implications for research, in *Time, Roles, and Self in Old Age,* p. 31.

[9]Augustine (1977) The confessions, in *The Philosophy of the Middle Ages* (A. Hyman and J. J. Walsh, eds.), Hackett Publishing Company, Indianapolis, IN, pp. 75–113, at 85.

[10]Augustine, The confessions, in *The Philosophy of the Middle Ages,* pp. 87,88 .

[11]J. J. C. Smart (1967) Time, in *The Encyclopedia of Philosophy* (P. Edwards, ed.) McMillan Publishing Company, New York, NY, pp. 126–134.

[12]G. J. Whitsow and A. Einstein (1967) in *The Encyclopedia of Philosophy* (P. Edwards, ed.), McMillan Publishing Company, New York, NY, pp. 468–471.

[13]K. S. Markides and L. A. Ray (1988) Change in subjective age among the elderly. *Comprehensive Gerontology* **2,** 11–15

[14]B. L. Neugarten (1979) Time, age and the life cycle. *American Journal of Psychiatry* **136,** 887–894.

[15]Hendricks and Hendricks, Concepts of time and temporal construction among the aged, with implications for research, in *Time, Roles, and Self in Old Age,* pp. 39–40.

[16]R. Kastenbaum (1966) On the meaning of temporality in later life. *Journal of Genetic Psychology* **109,** 9–25 at 20.

[17]R. Kastenbaum, V. Derbin, P. Sabatini, and S. Artt (1972) The ages of me: Toward personal and interpersonal definitions of functional aging. *International Journal of Aging and Human Development* **3,** 197–211.

[18]S. K. Baum (1983–1984) Age identification in the elderly: Some theoretical considerations. *International Journal of Aging and Human Development* **18,** 25–30.

[19]K. S. Markides and J. S. Boldt (1983) Change in subjective age among the elderly. *The Gerontologist* **23,** 422–427.

[20]M. W. Linn and K. Hunter (1979) Perception of age in the elderly. *Journal of Gerontology* **34,** 46–52.

[21]S. K. Baum and R. L. Boxley (1983) Age identification in the elderly. *The Gerontologist* **23,** 532–537.

[22]N. E. Strumpf (1987) Probing the temporal world of the elderly. *International Journal of Nursing Studies* **24,** 201–214.

[23]Ibid., p. 211.

[24]R. J. Hulbert and W. Lens (1988) Time and self-identity. *International Journal of Aging and Human Development* **27,** 293–303.

The Absurd

Thomas Nagel

Most people feel on occasion that life is absurd, and some feel it vividly and continually. Yet the reasons usually offered in defense of this conviction are patently inadequate: they *could* not really explain why life is absurd. Why then do they provide a natural expression for the sense that it is?

Minuteness and Brevity

Consider some examples. It is often remarked that nothing we do now will matter in a million years. If that is true, then by the same token, nothing that will be the case in a million years matters now. In particular, it does not matter now that in a million years nothing we do now will matter. Moreover, even if what we did now *were* going to matter in a million years, how could that keep our present concerns from being absurd? If their mattering now is not enough to accomplish that, how would it help if they mattered a million years from now?

Whether what we do now will matter in a million years could make the crucial difference only if its mattering in a million years depended on its mattering, period. However, then to deny that whatever happens now will matter in a million years is to beg the question against its mattering, period, for in that sense one cannot know that it will not matter in a million years whether (for example) someone now is happy or miserable, without knowing that it does not matter, period.

Aging and Ethics Ed.: N. Jecker

What we say to convey the absurdity of our lives often has to do with space or time: we are tiny specks in the infinite vastness of the universe; our lives are mere instants even on a geological time scale, let alone a cosmic one; we will all be dead any minute, but, of course, none of these evident facts can be what *makes* life absurd, if it is absurd. For suppose we lived forever; would not a life that is absurd if it lasts 70 years be infinitely absurd if it lasted through eternity? Also, if our lives are absurd given our present size, why would they be any less absurd if we filled the universe (either because we were larger or because the universe was smaller)? Reflection on our minuteness and brevity appears to be intimately connected with the sense that life is meaningless, but it is not clear what the connection is.

Another inadequate argument is that, because we are going to die, all chains of justification must leave off in midair: one studies and works to earn money to pay for clothing, housing, entertainment, food, to sustain oneself from year to year, and perhaps to support a family and pursue a career—but to what final end? All of it is an elaborate journey leading nowhere. (One will also have some effect on other people's lives, but that simply reproduces the problem, for they will die, too.)

There are several replies to this argument. First, life does not consist of a sequence of activities, each of which has as its purpose some later member of the sequence. Chains of justification come repeatedly to an end within life, and whether the process as a whole can be justified has no bearing on the finality of these end-points. No further justification is needed to make it reasonable to take aspirin for a headache, attend an exhibit of the work of a painter one admires, or stop a child from putting his or her hand on a hot stove. No larger context or further purpose is needed to prevent these acts from being pointless.

Even if someone wished to supply a further justification for pursuing all the things in life that are commonly regarded as self-justifying, that justification would have to end somewhere, too. If *nothing* can justify unless it is justified in terms of something outside itself, which is also justified, then an infinite regress results, and no chain of justification can be complete. Moreover, if a finite chain of reasons cannot justify anything, what could be accomplished by an infinite chain, each link of which must be justified by something outside itself?

Since justifications must come to an end somewhere, nothing is gained by denying that they end where they appear to, within life—or by trying to subsume the multiple, often trivial ordinary justifications of action under a single, controlling life scheme. We can be satisfied more easily than that. In fact, through its misrepresentation of the process of justification, the argument makes a vacuous demand. It insists that the reasons available within life are incomplete, but suggests thereby that all reasons that come to an end are incomplete. This makes it impossible to supply any reasons at all.

The standard arguments for absurdity appear, therefore, to fail as arguments. Yet I believe they attempt to express something that is difficult to state, but fundamentally correct.

Seriousness and Doubt

In ordinary life, a situation is absurd when it includes a conspicuous discrepancy between pretension or aspiration and reality: someone gives a complicated speech in support of a motion that has already been passed; a notorious criminal is made president of a major philanthropic foundation; you declare your love over the telephone to a recorded announcement; as you are being knighted, your pants fall down. When a person finds himself or herself in an absurd situation, he will usually attempt to change it by modifying his or her aspirations, trying to bring reality into better accord with them, or removing himself or herself from the situation entirely. We are not always willing or able to extricate ourselves from a position whose absurdity has become clear to us. Nevertheless, it is usually possible to imagine some change that would remove the absurdity—whether or not we can or will implement it. The sense that life as a whole is absurd arises when we perceive, perhaps dimly, an inflated pretension or aspiration that is inseparable from the continuation of human life and that makes its absurdity inescapable, short of escape from life itself.

Many people's lives are absurd, temporarily or permanently, for conventional reasons having to do with their particular ambitions, circumstances, and personal relations. If there is a philosophical sense of absurdity, however, it must arise from the perception of something universal—some respect in which pretension and reality inevitably clash for us all. This condition is supplied, I shall argue, by the collision between the seriousness with which we take our lives and the perpetual possibility of regarding everything about which we are serious as arbitrary, or open to doubt.

We cannot live human lives without energy and attention, nor without making choices that show that we take some things more seriously than others. Yet we have always available a point of view outside the particular form of our lives, from which the seriousness appears gratuitous. These two inescapable viewpoints collide in us, and that is what makes life absurd. It is absurd because we ignore the doubts that we know cannot be settled, continuing to live with nearly undiminished seriousness in spite of them. This analysis requires defense in two respects: first as regards the unavoidability of seriousness; second as regards the inescapability of doubt.

We take ourselves seriously whether we lead serious lives or not, and whether we are concerned primarily with fame, pleasure, virtue, luxury, triumph, beauty, justice, knowledge, salvation, or mere survival. If we take other people seriously and devote ourselves to them, that only multiplies the problem. Human life is full of effort, plans, calculation, success, and failure: we *pursue* our lives with varying degrees of sloth and energy.

It would be different if we could not step back and reflect on the process, but were merely led from impulse to impulse without self-consciousness, but human beings do not act solely on impulse. They are prudent, they reflect, they weigh consequences, and they ask whether what they are doing is worthwhile. Not only are their lives full of particular choices that hang together in larger activities with temporal structure, but they also decide in the broadest terms what to pursue and what to avoid, what the priorities among their various aims should be, and what kind of people they want to be or become. Some people are faced with such choices by the large decisions they make from time to time; some merely by reflection on the course their lives are taking as the product of countless small decisions. They decide whom to marry, what profession to follow, whether to join the Country Club or the Resistance, or they may just wonder why they go on being salespeople, academics, or taxi drivers, and then stop thinking about it after a certain period of inconclusive reflection.

Although they may be motivated from act to act by those immediate needs with which life presents them, they allow the process to continue by adhering to the general system of habits and the form of life in which such motives have their place—or perhaps only by clinging to life itself. They spend enormous quantities of energy, risk, and calculation on the details. Think of how an ordinary individual sweats over his or her appearance, health, sex life, emotional honesty, social utility, self-knowledge, the quality of his or her ties with family, colleagues, and friends, how well he or she does his or her job, and whether he or she understands the world and what is going on in it. Leading a human life is a full-time occupation, to which everyone devotes decades of intense concern.

This fact is so obvious that it is hard to find it extraordinary and important. Each of us lives his or her own life—lives with himself 24 hours a day. What else is one supposed to do—live someone else's life? Yet humans have the special capacity to step back and survey themselves, and the lives to which they are committed, with that detached amazement that comes from watching an ant struggle up a heap of sand. Without developing the illusion that they are able to escape from their highly specific and idiosyncratic position, they can view it *sub specie aeternitatis*—and the view is at once sobering and comical.

The crucial backward step is not taken by asking for still another justification in the chain, and failing to get it. The objections to that line of attack have already been stated; justifications come to an end, but this is precisely what provides universal doubt with its object. We step back to find that the whole system of justification and criticism, which controls our choices and supports our claims to rationality, rests on responses and habits that we never question, that we should not know how to defend without circularity, and to which we shall continue to adhere even after they are called into question.

The things we do or want without reasons, and without requiring reasons—the things that define what is a reason for us and what is not—are the starting points of our skepticism. We see ourselves from outside, and all the contingency and specificity of our aims and pursuits become clear. Yet when we take this view and recognize what we do as arbitrary, it does not disengage us from life, and there lies our absurdity: not in the fact that such an external view can be taken of us, but in the fact that we ourselves can take it, without ceasing to be the persons whose ultimate concerns are so coolly regarded.

Larger Purposes

One may try to escape the position by seeking broader ultimate concerns, from which it is impossible to step back—the idea being that absurdity results because what we take seriously is something small, insignificant, and individual. Those seeking to supply their lives with meaning usually envision a role or function in something larger than themselves. They therefore seek fulfillment in service to society, the state, the revolution, the progress of history, the advance of science, or religion and the glory of God.

However, a role in some larger enterprise cannot confer significance unless that enterprise is itself significant, and its significance must come back to what we can understand, or it will not even appear to give us what we are seeking. If we learned that we were being raised to provide food for other creatures fond of human flesh, who planned to turn us into cutlets before we got too stringy—even if we learned that the human race had been developed by animal breeders precisely for this purpose—that would still not give our lives meaning, for two reasons. First, we would still be in the dark as to the significance of the lives of those other beings; second, although we might acknowledge that this culinary role would make our lives meaningful to them, it is not clear how it would make them meaningful to us.

Admittedly, the usual form of service to a higher being is different from this. One is supposed to behold and partake of the glory of God, for example, in a way in which chickens do not share in the glory of coq au vin. The same is true of service to a state, a movement, or a revolution. People can come to feel, when they are part of something bigger, that it is part of them, too. They worry less about what is peculiar to themselves, but identify enough with the larger enterprise to find their role in it fulfilling.

However, any such larger purpose can be put in doubt in the same way that the aims of an individual life can be and for the same reasons. It is as legitimate to find ultimate justification there as to find it earlier, among the details of individual life, but this does not alter the fact that justifications come to an end when we are content to have them end—when we do not find it necessary to look any further. If we can step back from the purposes of individual life and doubt their point, we can step back also from the progress of human history, or of science, or the success of a society, or the kingdom, power, and glory of God,[1] and put all these things into question in the same way. What seems to us to confer meaning, justification, and significance does so in virtue of the fact that we need no more reasons after a certain point.

What makes doubt inescapable with regard to the limited aims of individual life also makes it inescapable with regard to any larger purpose that encourages the sense that life is meaningful. Once the fundamental doubt has begun, it cannot be laid to rest.

Camus maintains in *The Myth of Sisyphus* that the absurd arises because the world fails to meet our demands for meaning. This suggests that the world might satisfy those demands if it were different, but now we can see that this is not the case. There does not appear to be any conceivable world (containing us) about which unsettlable doubts could not arise. Consequently, the absurdity of our situation derives not from a collision between our expectations and the world, but from a collision within ourselves.

The Backward Step

It may be objected that the standpoint from which these doubts are supposed to be felt does not exist—that if we take the recommended backward step, we will land on thin air, without any basis for judgment about the natural responses we are supposed to be surveying. If we retain our usual standards of what is important, then questions about the significance of what we are doing with our lives will be answerable in the usual way. However, if we do not, then those questions can mean nothing to us, since there is no longer any content to the idea of what matters and, hence, no content to the idea that nothing does.

However, this objection misconceives the nature of the backward step. It is not supposed to give us an understanding of what is *really* important, so that we see by contrast that our lives are insignificant. We never, in the course of these reflections, abandon the ordinary standards that guide our lives. We merely observe them in operation, and recognize that if they are called into question we can justify them only by reference to themselves, uselessly. We adhere to them because of the way we are put together; what seems to us important, serious, or valuable would not seem so if we were differently constituted.

In ordinary life, to be sure, we do not judge a situation absurd unless we have in mind some standards of seriousness, significance, or harmony with which the absurd can be contrasted. This contrast is not implied by the philosophical judgment of absurdity, and that might be thought to make the concept unsuitable for the expression of such judgments. This is not so, however, for the philosophical judgment depends on another contrast that makes it a natural extension from more ordinary cases. It departs from them only in contrasting the pretensions of life with a larger context in which no standards can be discovered, rather than with a context from which alternative, overriding standards may be applied.

Philosophical Skepticism

In this respect, as in others, philosophical perception of the absurd resembles epistemological skepticism. In both cases, the final, philosophical doubt is not contrasted with any unchallenged certainties, though it is arrived at by extrapolation from examples of doubt within the system of evidence or justification, where a contrast with other certainties is implied. In both cases, our limitedness joins with a capacity to transcend those limitations in thought (thus seeing them as limitations and as inescapable).

Skepticism begins when we include ourselves in the world about which we claim knowledge. We notice that certain types of evidence convince us that we are content to allow justifications of belief to come to an end at certain points, and that we feel we know many things even without knowing or having grounds for believing the denial of others that, if true, would make what we claim to know false.

For example, I know that I am looking at a piece of paper, although I have no adequate grounds to claim I know that I am not dreaming, and if I am dreaming, then I am not looking at a piece of paper. Here, an ordinary conception of how appearance may diverge from reality is employed to show that we take our world largely for granted; the certainty that we are not dreaming cannot be justified except circularly, in terms of those very appearances that are being put in doubt. It is somewhat far-fetched to suggest I may be dreaming, but the possibility is only illustrative. It reveals that our claim to knowledge depends on our not feeling it necessary to exclude certain incompatible alternatives, and the dreaming possibility or the total-hallucination possibility is just a representative for limitless possibilities, most of which we cannot even conceive.[2]

Once we have taken the backward step to an abstract view of our whole system of beliefs, evidence, and justification, and seen that it works only, despite its pretensions, by taking the world largely for granted, we are *not* in a position to contrast all these appearances with an alternative reality. We cannot shed our ordinary responses, and if we could, it would leave us with no means of conceiving a reality of any kind.

It is the same in the practical domain. We do not step outside our lives to a new vantage point from which we see what is really, objectively significant. We continue to take life largely for granted, although seeing that all our decisions and certainties are possible only because there is a great deal we do not bother to rule out.

Both epistemological skepticism and a sense of the absurd can be reached via initial doubts posed within systems of evidence and justification that we accept, and can be stated without violence to our ordinary concepts. We can ask not only why we should believe there is a floor under us, but also why we should believe the evidence of our senses at all—and at some point, the framable questions will have outlasted the answers. Similarly, we can ask not only why we should take aspirin, but why we should take trouble over our own comfort at all. The fact that we shall take the aspirin without waiting for an answer to this last question does not show that it is an unreal question. We shall also continue to believe there is a floor under us without waiting for an answer to the other question. In both cases, it is this unsupported natural confidence that generates skeptical doubts; so it cannot be used to settle them.

Philosophical skepticism does not cause us to abandon our ordinary beliefs, but it lends them a peculiar flavor. After acknowledging that their truth is incompatible with possibilities that we have no grounds for believing do not obtain—apart from grounds in those very beliefs that we have called into question—we return to our familiar convictions with a certain irony and resignation. Unable to abandon the natural responses on which they depend, we take them back, like a spouse who has run off with someone else and then decided to return, but we regard them differently (not that the new attitude is necessarily inferior to the old, in either case).

The same situation obtains after we have put in question the seriousness with which we take our lives and human life in general and have looked at ourselves without presuppositions. We then return to our lives, as we must, but our seriousness is laced with irony. Not that irony enables us to escape the absurd. It is useless to mutter: "Life is meaningless; life is meaningless..." as an accompaniment to everything we do. In continuing to live, work, and strive, we take ourselves seriously in action no matter what we say.

What sustains us, in belief as in action, is not reason or justification, but something more basic than these—for we go on in the same way even after we are convinced that the reasons have given out.[3] If we tried to rely entirely on reason, and pressed it hard, our lives and beliefs would collapse—a form of madness that may actually occur if the inertial force of taking the world and life for granted is somehow lost. If we lose our grip on that, reason will not give it back to us.

Self-Consciousness and Absurdity

In viewing ourselves from a perspective broader than we can occupy in the flesh, we become spectators of our own lives. We cannot do very much as pure spectators of our own lives, so we continue to lead them, and devote ourselves to what we are able at the same time to view as no more than a curiosity, like the ritual of an alien religion.

This explains why the sense of absurdity finds its natural expression in those bad arguments with which the discussion began. References to our small size and short lifespan, and to the fact that all humans will eventually vanish without a trace are metaphors for the backward step that permits us to regard ourselves from without and to find the particular form of our lives curious and slightly surprising. By feigning a nebula's eye view, we illustrate the capacity to see ourselves without presuppositions, as arbitrary, idiosyncratic, highly specific occupants of the world, one of countless possible forms of life.

Before turning to the question whether the absurdity of our lives is something to be regretted and, if possible, escaped, let me consider what would have to be given up in order to avoid it. Why is the life of a mouse not absurd? The orbit of the moon is not absurd either, but that involves no strivings or aims at all. A mouse, however, has to work to stay alive. Yet he is not absurd, because he lacks the capacities for self-consciousness and self-transcendence that would enable him to see that he is only a mouse. If that *did* happen, his life would become absurd, since self-awareness would not make him cease to be a mouse and would not enable him to rise above his mousely strivings. Bringing his newfound self-consciousness with him, he would have to return to his meager yet frantic life, full of doubts that he was unable to answer, but also full of purposes that he was unable to abandon.

Given that the transcendental step is natural to us humans, can we avoid absurdity by refusing to take that step and remaining entirely within our sublunar lives? Well, we cannot refuse consciously, for to do that we would have to be aware of the viewpoint we were refusing to adopt. The only way to avoid the relevant self-consciousness would be either never to attain it or to forget it—neither of which can be achieved by the will.

On the other hand, it is possible to expend effort on an attempt to destroy the other component of the absurd—abandoning one's earthly, individual, human life in order to identify as completely as possible with that universal viewpoint from which human life seems arbitrary and trivial. (This appears to be the ideal of certain Oriental religions.) If one succeeds, then one will not have to drag the superior awareness through a strenuous mundane life, and absurdity will be diminished.

However, insofar as this self-etiolation is the result of effort, willpower, asceticism, and so forth, it requires that one take oneself seriously as an individual—that one be willing to take considerable trouble to avoid being creaturely and absurd. Thus, one may undermine the aim of unworldliness by pursuing it too vigorously. Still, if someone simply allowed his or her individual, animal nature to drift and respond to impulse, without making the pursuit of its needs a central conscious aim, then he or she might, at considerable dissociative cost, achieve a life that was less absurd than most. It would not be a meaningful life either, of course, but it would not involve the engagement of a transcendent awareness in the assiduous pursuit of mundane goals. Also, that is the main condition of absurdity—the dragooning of an unconvinced transcendent consciousness into the service of an immanent, limited enterprise like a human life.

Insight and Absurdity

The final escape is suicide, but before adopting any hasty solutions, it would be wise to consider carefully whether the absurdity of our existence truly presents us with a *problem,* to which some solution must be found—a way of dealing with *prima facie* disaster. That is certainly the attitude with which Camus approaches the issue, and it gains support from the fact that we are all eager to escape from absurd situations on a smaller scale.

Camus—not on uniformly good grounds—rejects suicide and the other solutions he regards as escapist. What he recommends is defiance or scorn. We can salvage our dignity, he appears to believe, by shaking a fist at the world that is deaf to our pleas, and continuing to live in spite of it. This will not make our lives unabsurd, but it will lend them a certain nobility.[4]

This seems to me romantic and slightly self-pitying. Our absurdity warrants neither that much distress nor that much defiance. At the risk of falling into romanticism by a different route, I would argue that absurdity is one of the most human things about us: a manifestation of our most advanced and interesting characteristics. Like skepticism in epistemology, it is possible only because we possess a certain kind of insight—the capacity to transcend ourselves in thought.

If a sense of the absurd is a way of perceiving our true situation (even though the situation is not absurd until the perception arises), then what reason can we have to resent or escape it? Like the capacity for epistemological skepticism, it results from the ability to understand our human limitations. It need not be a matter for agony unless we make it so. Nor need it evoke a defiant contempt of fate that allows us to feel brave or proud. Such dramatics, even if carried on in private, betray a failure to appreciate the cosmic unimportance of the situation. If *sub specie aeternitatis* there is no reason to believe that anything matters, then that does not matter either, and we can approach our absurd lives with irony instead of heroism or despair.

Notes and References

[1]Cf. R. Nozick (1971) Teleology, *Mosaic* **XII,** I, 27,28.

[2]I am aware that skepticism about the external world is widely thought to have been refuted, but I have remained convinced of its irrefutability since being exposed at Berkeley to Thompson Clarke's largely unpublished ideas on the subject.

[3]As Hume says in a famous passage of the *Treatise:* "Most fortunately it happens, that since reason is incapable of dispelling these clouds, nature herself suffices to that purpose, and cure me of this philosophical melancholy and delirium, either by relaxing this bent of mind, or by some avocation, and lively impression of my senses, which obliterate all these chimeras. I dine, I play a game of backgammon, I converse, and am merry with my friends; and when after three or four hours' amusement, I would return to these speculations, they appear so cold, and strained, and ridiculous, that I cannot find in my heart to enter into them any farther" (Book 1, Part 4, Section 7; Selby-Bigge, p. 269).

[4]"Sisyphus, proletarian of the gods, powerless and rebellious, knows the whole extent of his wretched condition: it is what he thinks of during his descent. The lucidity that was to constitute his torture at the same time crowns his victory. There is no fate that cannot be surmounted by scorn" (*The Myth of Sisyphus,* Vintage edition, p. 90).

Index of Names

Aeschylus, 356
Albee, Edward, 114, 117
Aries, Philippe, 137
Aristophanes, 356
Aristotle, 56, 59, 60, 75, 97, 367, 368, 370
Auden, W. D., 83
St. Augustine, 36, 69–72, 368–370
Avron, Jerome, 221

Beckett, Samuel, 54
Berger, Peter, 77
Binstock, Robert, 326
Bloch, Ernst, 334
Blum, Lawrence, 199, 200
Blythe, Ronald, 113, 116, 119, 124
Brody, Howard, 203
Broome, John, 288, 290
Burt, Robert, 199

Callahan, Daniel, 239, 248, 253–257, 259–261, 264, 265, 352, 363, 273–276, 280, 281, 321, 323, 327
Camus, Albert, 382, 387
Childress, James, 324
Churchill, Larry, 248, 262–265
Cicero, Marcus Tullius, 60, 97
Copernicus, Nicolaus, 4

Daniels, Norman, 248–257, 270–272, 274, 276, 278, 279
de Beauvoir, Simone, 63, 113
Descartes, Rene, 276
Dostoyevsky, Feodor, 38

Einstein, Albert, 369, 370
Engelhardt, Tristram, 252
Erikson, Erik, 58, 60, 61, 96
Euripides, 356

Freud, Sigmund, 60, 61, 69, 99, 100
Friedan, Betty, 135

Gadow, Sally, 203

Hegel, Georg Wilhelm Friedrich, 57
Heidegger, Martin, 65, 190–192
Hippocrates, 97
Hume, David, 341–343, 345, 348, 350

Jaspers, Karl, 54, 65
Jecker, Nancy, 323
Jonsen, Albert, 332
Jung, Carl, 96

Kafka, Franz, 325
Kant, Immanuel, 32, 57, 65
Kapp, Marshall, 333
Kastenbaum, Robert, 370
Kierkegaard, Soren, 72

Lamm, Richard, 248–250

Lasch, Christopher, 53
Leibniz, Gotffried Wilhelm, 9
Lifton, Robert, 62, 78

MacIntyre, Alasdaire, 342
Mann, Thomas, 101
Mill, John Stuart, 249
Miller, Arthur, 356
Mishan, E, 288
Montaigne, Michel, 69, 72, 351

Newton, Issac, 7, 18, 19, 368
Nietzsche, Friedrich, 70, 115, 343

Parfit, Derek, 252
Plato, 59, 188
Preston, Samuel, 129, 220, 221

Rawls, John, 235, 250, 252, 253, 347, 348
Rhoden, Nancy, 199, 203, 206
Roscowe, Irving, 96, 202
Rousseau, Jean Jacques, 69, 71, 72
Rossi, Alice, 142
Royce, Josiah, 189, 191
Ruddick, Sara, 199, 200, 212
Russell, Louise, 293

Sartre, Jean Paul, 71, 116
Scheler, Max, 343, 344, 346, 351
Schoeman, Ferdinand, 206
Sexton, Anne, 108
Shakespeare, William, 97
Smart, J. C. C., 21
Smith, Adam, 261, 262
Socrates, 7, 8, 59, 60
Sontag, Susan, 325
Sophocles, 99–102, 105–107, 356
Spinoza, Baruch, 56

Thoreau, Henry David, 213, 275–281
Tolstoy, Leo, 3, 38

Voltaire, 53

Yeats, William Butler, 80

Walzer, Michael, 348
Williams, Tennessee, 363
Wittgenstein, Ludwig, 21, 55, 70

Index of Subjects

Ages of life, 97, 99, 100, 105, 107
Ages of man, *see* Ages of life; Joumey of life
Allocation, *see* Rationing
Alzheimer's disease, 94, 135, 228, 316
Autonomy, 81
  best interest versus, 180–184, 318
  capacity for, 173–179, 192, 200, 203–206
  decisional autonomy, 177, 186
  definition of, 314
  executional autonomy, 177, 186
  external constraints, 174, 178–180
  family autonomy, 206, 207, 209
  free choice, 157, 165, 166, 314, 319
  internal constraints, 174–179
  intimacy versus, 199–216
  limits of, 172–175, 177, 192, 210, 211, 252
  and moral philosophy, 200
  principle of, 200
  self-reliance and, 276

Body
  aging of, 113, 114, 116, 117
  alienation from, 113, 114, 117, 118, 190
  consciousness of, 113
  culture and, 116, 118
  death and, 113, 190
  in illness, 115
  meanings and, 114–116
  science and, 116
  self and, 113–116, 118, 119
  sexuality and, 118
  technology and, 156, 190
  value and, 114, 115
  youth and, 114, 115, 117

Caregiving, 125, 126, 130, 132–135, 141, 155–170, 184, 185, 196–198, 201, 205, 209, 211, 231, 232, 330, 350, 357, 358, 319–322

Death, *see* Meaning of life, death and; Meaning of old age, death and

Ethics, *see* Meaning of life, value and; Family, ethics in
Euthanasia, 41, 58
Existential, 55, 60, 63–66, 75, 85, 106, 107, 117
Existentialism, *see* Existential

Families (*see also* Women, in family; Justice, within family; Caregiving)
  abuse and, 321

death in, 93, 94, 188–194, 198, 265, 325
death of parent, 93–95, 106, 108, 124, 125, 128
definition of, 208, 209
demographics changes, 123–144
economics and, 132, 135–138, 140, 141, 151–153, 168, 196, 201, 202, 208, 231, 232, 263
ethics in, 131–135, 147–154, 155–170, 174–178, 184, 185, 189, 201–206, 212, 320
intimacy in, 93, 199–216
loyalty in, 187–198
medical decision making, role in, 174–178, 187–216, 232
men in, 129, 130, 134, 137, 138, 140–143

Generations (*see also* Women, mother–daughter relationship)
conflict between, 100, 133, 195, 220–222, 226, 230–233, 240, 241, 249, 258, 263, 314, 322, 364
relationships between, 126–129, 131–140, 142–144, 147, 151–154, 163, 200–204, 255, 264, 328
succession of, 93, 100, 106, 140, 257, 258, 359

Happiness, 1–3, 38–41, 58–60, 68, 69, 157, 220

Journey of life, 98, 99, 108

Justice (*see also* Meaning of life, values and; Family, ethics in)
contract theory and, 188
impartiality and, 188, 189, 212, 213
resentment and, 342–347, 349–351
sympathy and, 261–264, 362, 363
within family, 209–213

Life stages, 52, 57, 58, 60, 61, 65, 72, 75, 76, 78, 81, 82, 225, 237–239, 241, 242, 250, 252, 278, 279, 344, 371

Meaning of life (*see also* Meaning of old age; Time, meaning and)
awe and, 21, 22
Christianity and, 3–7, 10–12, 26, 29, 30, 32–42, 44–46
death and, 33, 43, 190–191, 225, 228, 376
existence and, 9–11, 20, 22, 23, 26
explanations, kinds of, 8–11, 15, 26, 27
explanations, limits of, 19, 20, 376, 377, 380–385
meaning of games and, 13–15
meaninglessness, 3, 6, 29, 31, 32, 46, 66–68, 375–378, 380–382
perspective and, 378–382, 384–386
science and, 4–7, 9, 12, 17, 19, 26–33, 44–46

values and, 37, 41, 42, 44, 45, 167
Meaning of old age
activity and, 52, 59, 62, 85, 191, 223, 255, 264, 360
aging society and, 54, 55
autobiography, 11, 55, 66–73, 253, 264, 265, 280
Christianity and, 61
community and, 107
contemplation and, 52, 62–64, 79, 85
culture and, 51–54, 62, 64, 74, 77, 78, 80, 82, 83, 95, 96, 97, 99, 107, 108, 223, 264, 265, 279, 324, 346, 347, 363
death and, 54, 60, 62, 64, 67, 73–75, 79, 80, 82, 104–106, 187, 191, 193, 195, 223–226, 232, 255, 256, 260, 265, 273, 274, 279, 280, 281, 322–324, 346, 347, 357–359, 363
development and, 52, 53, 57, 58, 59, 72, 74, 82–84, 372
future and, 55, 190, 191, 357, 371, 372
honor and, 346–350, 357, 362, 371, 372
liberation and, 191, 193, 223
meaning, levels of, 53, 54, 56, 57, 62, 73
meaninglessness, 64, 65, 70, 76, 77, 224, 256, 281
medicine and, 75, 76, 95, 196, 223–226, 228, 255, 256, 273, 357, 359, 360
narcissism and, 53, 54, 61, 62, 255, 323
transcendence, 52, 61–63, 75, 80, 95
values and, 344, 346
whole of life, 51, 52, 54–57, 66, 69, 83, 96, 99, 190, 192, 371
wisdom and, 73–75, 61, 69, 104, 106, 108, 195, 196
Medicine (*see also* Euthanasia; Families, role in medical decision making; Nurses; Rationing)
economics and, 188, 194, 210, 211, 219–223, 225–245, 247–260, 262, 263, 270, 271, 279, 280, 285, 304, 307–310, 313, 322–328, 330, 332, 341, 350, 359, 360, 362, 364
informed consent, 193, 204
living wills, 194, 195, 204, 205, 364
paternalism, 193
politics and, 222, 223, 230, 243, 307–334, 341, 350, 351, 364
progress and, 156, 188, 219, 228
prolonging life, 188, 193–197, 219, 222, 225, 226, 234, 237, 285–287, 289–301, 359
Mercy killing, *see* Euthanasia
Morality, *see* Meaning of life, values and; Families, ethics in

Nature
meanings of, 269, 270, 272–279, 281, 282

natural death, 270
natural lifespan, 224, 225, 253–255, 260, 270, 273–275, 280, 281
versus disease, 270–272
renewal of, 361
Nurses (*see also* Medicine
role in medical decision making, 171–186

Prudential lifespan account, 237–245, 250, 251

Rationing (*see also* Families, economics and; Medicine, economics and)
by age, 219–222, 225, 226, 229, 232, 236–240, 244, 245, 248–251, 253–263, 269–274, 278, 281, 286, 287, 289, 301–303, 322–327, 329, 330
effects on vulnerable, 327, 328, 330–333
effects on women, 327, 330
of high technology medical care, 187, 188, 195, 219, 220, 229, 230, 234, 236, 242–244, 247
of life-extending medical care, 187, 195
under Medicare, 222, 229, 315
soft rationing, 210
standards for, 204–206, 235, 308, 323, 331
Retirement, 52, 77, 79, 80, 138, 191

Self-deception, 16, 65, 67, 193
Surrogate decision making, *see* Family, role in medical decision making; Nurses, role in medical decision making; Medicine, living wills
Stages of life, *see* Life stages
Stoicism, 60, 61, 69
Suicide, 1–3, 38–41, 58, 60, 171, 195, 358, 387

Temporality, *see* Time; Meaning of old age, future and
Time (*see also* Meaning of old age, future and)
definitions of, 367–369
human time, 73, 97
meaning and, 23, 77, 85, 106, 162, 278, 361, 367, 376
objective time, 368–372
subjective time, 368, 369, 371, 372
timeliness and, 85, 124, 302, 358, 369, 370
values and, 367, 370, 372

Utilitarianism, 248–251, 342, 362

Women (*see also* Rationing, effects on women)
divorce and, 137–140, 142
in family, 129, 130, 133, 134, 139–143, 166, 320, 321
longevity and, 130, 320
mother–daughter relationship, 133, 134, 140, 141
self-sacrifice and, 320, 321, 324
superwoman, 135
widowed, 130, 220
work and, 134, 135, 320